Leckie
the education publisher
for Scotland

T0173333

National 5
GEOGRAPHY

Revision + Practice
2 Books in 1

ISBN 9780008435394

Published by
Leckie
An imprint of HarperCollinsPublishers
Westerhill Road, Bishopbriggs, Glasgow, G64 2QT
T: 0844 576 8126 F: 0844 576 8131
leckiescotland@harpercollins.co.uk www.leckiescotland.co.uk

HarperCollins Publishers
Macken House, 39/40 Mayor Street Upper, Dublin 1, Ireland, D01 C9W8

Publisher: Sarah Mitchell
Project Managers: Harley Griffi ths and Lauren Murray

Special thanks to
QBS (layout and illustration)

Printed in the United Kingdom.

A CIP Catalogue record for this book is available from the British Library.

Acknowledgements
We would like to thank the following for permission to reproduce their material:
Photograph of Glenshee Ski Centre on page 36 © Brendan Howard / Shutterstock.com; Photograph of Weston-Super-Mare beach on page 45 © Mike Charles / Shutterstock.com; Photograph of Masai family on page 52 © Birute Vijeikiene / Shutterstock.com; Photograph of doctors' surgery in Kenya on page 57 © spirit of america / Shutterstock.com; Photograph of London 2012 Olympics on page 64 © pcruciatti / Shutterstock.com; Photograph of damage caused by Hurricane Sandy on page 78 © Leonard Zhukovsky / Shutterstock.com; Photograph of oil spill on page 84 © National Geographic / Getty Images; Photograph of Haitian health worker on page 101 © arindambanerjee / Shutterstock.com Photograph of doctors' surgery in the UK on page 124 © Dmitry Berkut / Shutterstock.com

All other images from Shutterstock or Thinkstock.

OS Maps
Ordnance Survey ® map data licensed with the permission of the Controller of Her Majesty's Stationery Office. © Crown copyright. Licence number 100018598.

Whilst every effort has been made to trace the copyright holders, in cases where this has been unsuccessful, or if any have inadvertently been overlooked, the Publishers would gladly receive any information enabling them to rectify any error or omission at the first opportunity.

ebook

To access the ebook version of this Revision Guide visit
www.collins.co.uk/ebooks
and follow the step-by-step instructions.

Contents

ANSWERS Check your answers to the practice test papers online:
www.collins.co.uk/pages/Scottish-curriculum-free-resources

Introduction

Complete Revision and Practice

This Complete **two-in-one Revision and Practice** book is designed to support you as students of National 5 Geography. It can be used either in the classroom, for regular study and homework, or for exam revision. By combining **a revision guide and two full sets of practice test papers**, this book includes everything you need to be fully familiar with the National 5 Geography exam. As well as including ALL the core course content with practice opportunities, there is a glossary and comprehensive assignment and exam preparation advice with revision question and practice test paper answers provided online at www.collins.co.uk/pages/Scottish-curriculum-free-resources.

Course Content and Assignment

In the National 5 Geography course you will study a wide variety of topics, which will be broken down into three sections: Physical Environments, Human Environments and Global Issues.

You will also complete an Assignment, which may have taken the form of a project, a presentation, or a report. Throughout this book you will find some suggestions of what you could do to complete this component of the National 5 Geography Award.

Key skills in Geography include being able to use a range of numerical and graphical information in the context of physical, human and global geographical issues, drawing on detailed knowledge and understanding of a significant global geographical issue and using skills to research in depth a specific line of enquiry. You should also familiarise yourself with Ordnance Survey (OS) maps – a key tool for Geography students in learning about the world.

Section 1: Physical Environments

In this section you study weather and climate, and then choose between rivers and limestone, or glaciation and coasts. Alternatively, you may have studied all four topics in preparation for moving onto Higher Geography at a later date. You will study the key processes, the formation of features and a case study of a named coastal area, such as the Dorset Coast or the Pembrokshire Coast, or upland area, such as the Cairngorms or the Lake District, and then apply this to examination questions.

In the exam you will have a choice of questions for Section 1: Physical Environments. You should attempt either Question 1 or Question 2, then answer the rest of the questions in this section. This will account for 30 marks.

Fig i. A limestone pavement in Malham

Section 2: Human Environments

In this section you will study contrasts between developed and developing countries, in both urban and rural areas. Again you should have named case study material, having looked at specific named countries and cities from across the globe. You will also learn about world population change in a variety of worldwide contexts.

In the exam you will have to attempt all questions in section 2: Human Environments. This too will account for 30 marks.

Fig ii. A Brazilian favela

Section 3: Global Issues

This unit is set on a global scale so you can use case studies from any country in the world. You should study at least two global issues from the following topics:

- Climate change
- Natural regions
- Environmental hazards
- Trade and globalisation
- Tourism
- Development and health

In the exam you will have to attempt 2 questions in section 3: Global Issues. This final section will account for 20 marks.

Fig iii. A crowded tourist beach

In summary, the National 5 Geography examination question paper is scored out of a total of 80 marks and contributes to 80% of your overall mark.

Assignment project ideas

A number of project ideas are included in this book as well as one or two more detailed guides to help you work on your own assignments. Here are some suggestions that will help you to design your own project.

1. Carry out a study in one of the key sectors of a city you live in or know well, e.g. investigate changes in Edinburgh's central business district (C.B.D.).

    *   Gather your data, process it, present it, analyse it.
    *   Key skills required here include: land use mapping, surveys (traffic, pedestrians, building height), questionnaires, data processing, data analysis.
    *   If you can carry out field work locally, you could use secondary data to compare how the city/town has changed over time. Use old OS maps and Google Earth to gather data.

Fig iv. Princes Street, Edinburgh

2. Physical landscape projects might involve mapping features seen in the field, drawing and annotating field sketches, drawing cross sections and explanatory diagrams, creating maps using Google satellite images as a base. Projects should try to explain how features form. Possible topics that might be investigated are:

    *   Rivers – What changes would you find along a river from source to mouth? How does the bed load change along a river's course? What changes in land use are there along a river's course?
    *   Limestone landscapes – Investigation into land use conflicts within the Yorkshire Dales National Park. An investigation into the advantages and disadvantages of a proposed development with the National Park, e.g. a quarry.
    *   Coasts – Investigation into the possible impact of a proposed coastal development and how this would affect the natural processes at work along the coastline. What are the advantages and disadvantages of a new coastal protection scheme?

3. Global issues can be investigated using a study question or a series of questions.

    *   What will the impact of a wind farm be on the local area?
    *   An environmental issue for and against: what are the merits of both sides of a debate and what conclusion do you arrive at?
    *   Why should more areas of Scotland be designated as national parks?
    *   Why do people object to developments in an environmentally sensitive area? (Local, UK, European Union (EU) or global.)

4. What are the effects of tourism on an area? Visitor survey, Internet research, relating the area to the Butler Model, environmental impact survey, debate.

5. A case study of a disease and how it affects people. Mapping distributions, diagrams defining transmission and environmental factors involved, economic and social effects, research, prevention and cure.

6. A comparison of a developed and developing country using statistical information from library and Internet sources.

7. A study of the impact of a development, e.g. hydroelectric power (HEP) scheme, industrial development on a fragile environment such as in the Tundra or in the Equatorial Rainforest.

8. A comparison of the trading patterns of particular developed and developing countries.

Fig v. Deforestation in the Amazon rainforest

In summary, the National 5 Geography assignment is scored out of a total of 20 marks and contributes to 20% of your overall mark.

Leckie
the education publisher
for Scotland

National 5
GEOGRAPHY

Revision Guide

Rob Hands, Alison Hughes
and Samantha Peck

Britain's weather

High and low pressure

Britain's weather is dominated by **Atlantic low pressure systems** (depressions or cyclones) that sweep across the British Isles from west to east. These bring a mix of warm **tropical maritime** air and cold **polar maritime** air. Cold air is denser so it moves more quickly and undercuts warmer air at the rear of the depression, or blocks its movement in at the leading edge of this feature (see Fig 1.1).

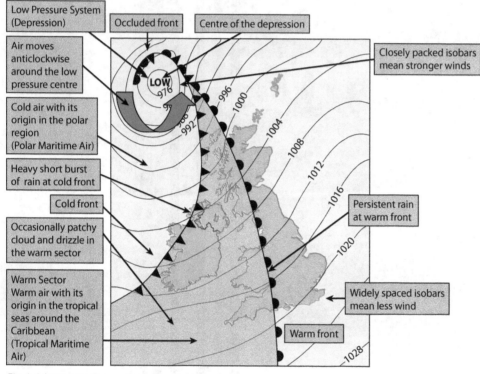

Low Pressure System (Depression)

Occluded front

Centre of the depression

Closely packed isobars mean stronger winds

Air moves anticlockwise around the low pressure centre

Cold air with its origin in the polar region (Polar Maritime Air)

Heavy short burst of rain at cold front

Cold front

Occasionally patchy cloud and drizzle in the warm sector

Warm Sector Warm air with its origin in the tropical seas around the Caribbean (Tropical Maritime Air)

Persistent rain at warm front

Widely spaced isobars mean less wind

Warm front

Fig 1.1 Low pressure system

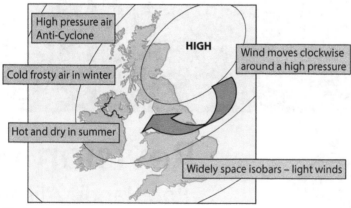

High pressure air Anti-Cyclone

HIGH

Wind moves clockwise around a high pressure

Cold frosty air in winter

Hot and dry in summer

Widely space isobars – light winds

Fig 1.2 High pressure system

What are weather fronts?

An air mass is a large body of air that has similar moisture, density and temperature characteristics. Air masses with different temperature, moisture and pressure characteristics tied up in depressions do not mix easily. A front is the boundary separating two air masses.

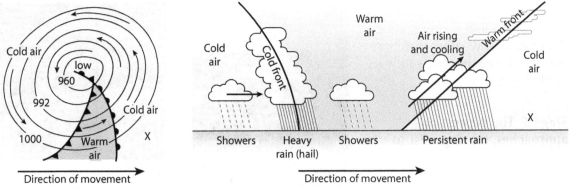

Fig 1.3 Where air masses meet

A warm front is where warm air contacts cold air. The contact zone is at ground level and reaches high above, up into the atmosphere. The diagram (Fig 1.3) demonstrates how at the warm front the warm sector is being forced to rise over the cold air in front of it. As it rises, it cools and condenses and forms clouds, leading to persistent rainfall on the ground. Behind the front, the warm sector is being squeezed and forced to rise by the heavier dense cold air driving in behind it. Here, showers of rain may occur.

A cold front is where cold air contacts warm air. At the cold front, heavy cloud is created by the associated uplift of warm air and a burst of heavy rain is likely. Behind the cold front, the polar maritime cold air is more stable and drier colder conditions might be expected. If the warm sector loses contact with the ground, the front becomes an **occluded** front, which will likely deliver heavy rain.

On weather maps, weather systems can be identified, along with their fronts and pressure readings. Atmospheric pressure is shown on the maps using contour-like **isobars.** These show pressure in **millibars** (mb) at intervals of 4 mb. Wind direction can be interpreted from the isobar pattern on a map. Winds blow **anti-clockwise** around low pressure systems just off parallel to the isobars in towards the centre of the low pressure system. Where isobars are close together this indicates a steep pressure gradient and therefore strong winds. Widely spaced isobars such as in Fig 1.2 mean little wind is being experienced on the ground.

Quick Test

1. What air masses are typically mixed in an Atlantic depression?
2. Which moves the quickest in a depression warm air or cold air?
3. What is a front?
4. What is created if the warm sector leaves the ground?
5. If winds move anti-clockwise around low pressure systems, which way do they move around a high pressure system?

The passage of a depression over Britain

Cyclonic systems (depressions and fronts) bring a sequence of weather events as they pass over Britain. Standing at any point on the ground, a weather watcher will see the sequence as follows: Study Fig 1.1

Fig 1.4 Dense cloud cover during a depression

Stage 1: The depression approaches, moving west to east.

Weather conditions: In the forward cold sector the weather is bright and sunny. High-level cloud (cirrus and cirro-stratus) begins to develop. A halo around the moon or sun can be observed as the warm front approaches.

Wind: Winds are light and south-westerly but begin to increase in strength.

Stage 2: The warm front arrives.

Weather conditions: At the warm front, there are long periods of persistent rain. There is dense cloud cover (stratus or strato-cumulus clouds) that gives full sky cover.

Wind: Wind speeds increase and the wind changes direction at the front, switching to westerly. Wind speed increases and may reach storm- or gale-force in a typical storm.

Stage 3: The warm sector.

Weather conditions: The rain eases and temperatures rise as the warm air passes over. It may become showery or drizzly.

Wind: Winds remain strong and westerly.

Stage 4: The cold front arrives.

Weather conditions: There is a heavy burst of rain for a short time as the steep front passes. Temperature drops as the cold air follows behind its front.

Wind: Winds back around from the north-west as the front passes, bringing in colder air.

Stage 5: Behind the cold front.

Weather conditions: The polar maritime or Arctic air is colder and stable so produces cold dry conditions with decreasing cloud (cumulus) as the weather clears up.

Wind: Winds from the north-west are blustery at first but wind speed decreases as the front moves away. Calmer conditions follow.

High pressure systems: anticyclones

Fig 1.5 Edinburgh in a high pressure system

High pressure systems are called anticyclones. They are less common in Britain but can build to bring long periods of stable weather. In summer, high pressure brings hot, dry conditions with low wind speeds while winter days are often cold and frosty, and fog and ice may develop. Night time temperatures are particularly low, and bright sunny winter days may stay below freezing. Easterly winds bring snow to eastern Britain in winter.

EXAM TIP

The best way to learn about weather is to follow weather forecasts on TV and the Internet. Search for the Met Office, BBC Weather Service or the Mountain Weather Information Service.

DID YOU KNOW?

The lowest recorded temperature in Scotland was −27°, first recorded in Braemar in 1982 and then again in the Highlands in 2010.

Weather plots

On complex weather maps and synoptic charts, codes are used to plot the weather at a weather station (for example at an airport). These provide a summary of the weather at a glance provided that you can work out the key that cracks the code.

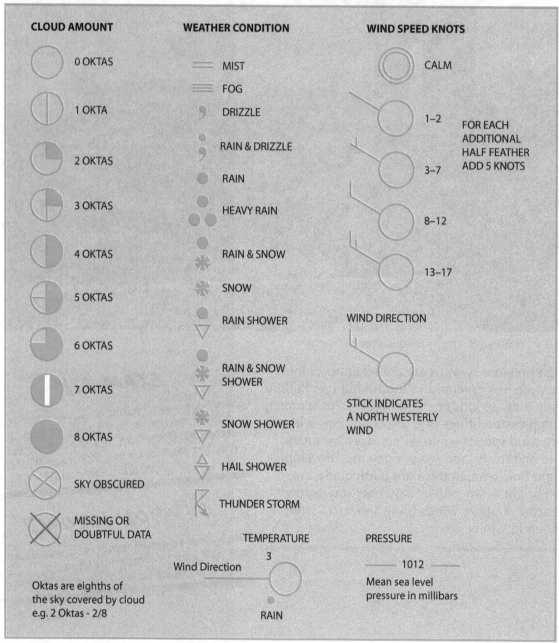

Fig 1.6 Synoptic chart symbols

Activity

1. Look at the three weather plots below and summarise the weather being experienced at each, using the weather plot symbols above to help you.

2. Draw weather plots showing the following weather conditions:

   a. The wind is northerly, blowing at 15 knots. The temperature is –6°C and there are snow showers. There are 7 oktas of cloud.

   b. The sky is obscured. There is fog. There is a light wind blowing from the south-east at 2 knots. The temperature is 4°C.

   c. The wind is south-westerly blowing at 40 knots. The sky is completely covered by cloud and there is rain and drizzle.

3. Study the weather map to the right.

   a. Summarise the weather at each of the weather stations a, b, e and g.

   b. Describe the weather at station h and explain why the weather is like that.

WEATHER CHART

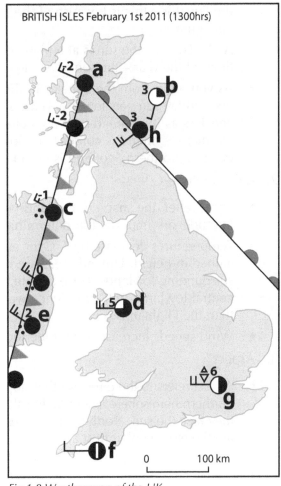

Fig 1.8 Weather map of the UK

Quick Test

1. In which direction do most low pressure systems (depressions) cross the British Isles? Choose from east to west; north to south; south-east to north-west; or west to east.

2. What would a weather observer notice as the cold front arrives?

3. What is another name for a high pressure system?

4. What weather conditions are likely with a high pressure system over Scotland in winter?

Factors affecting climate and weather in the UK

The climate and weather of a place depends on a number of factors:

1. Latitude

   - Temperatures generally decrease with distance from the Equator – this is because of the Earth's tilt and its curvature.

   - At the Equator, the sun is at a high angle at midday at all times of the year – the sun's heating effect is at its greatest.

   - As you move from the Equator to the poles, the angle of the midday sun gets lower and the sun's heating effect gets less as it is spread over a greater area. For example, Northern Scotland is colder than Southern England because northern Scotland is closer to the North Pole.

2. Relief/altitude

   - Local relief (the shape of the land) and the height (altitude) are important in determining the local climate.

   - Temperatures decrease by 1°C for every 100 metres gained in height. Upland areas are therefore cooler than surrounding lowlands. For example, in Scotland the central lowlands will experience warmer temperatures than the Highlands.

   - Wind speeds increase as altitude increases.

3. Aspect

   - Aspect refers to the direction that slopes are facing. In the northern hemisphere, south-facing slopes are warm as they face the sun. North-facing slopes are shaded and are therefore cooler: frost and snow lie longer in the winter and the slope does not get so hot in summer.

   - Slopes alter the angle at which the sun's rays hit the ground.

   - Land use can be affected by aspect – sunny south-facing slopes are better for agriculture.

4. Distance from the sea

   - Land and sea differ in their ability to absorb, transfer and reflect heat.

   - The sea can absorb heat to a greater depth (10 metres) than the land. This means it takes longer to heat up and longer to cool down.

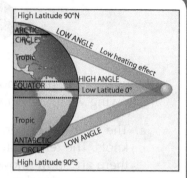

Fig 1.9 The effects of latitude

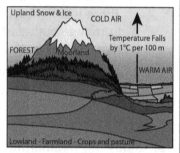

Fig 1.10 The effects of relief

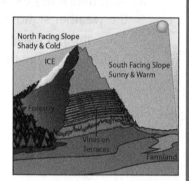

Fig 1.11 The effects of aspect

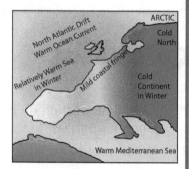

Fig 1.12 The effects of distance from the sea

- Places closer to the sea have milder winters and cooler summers. Oceans act as 'thermal reservoirs' (hot water bottles to us!) and so coastal areas rarely suffer from frost and heavy snowfall and have a lower annual range of temperature than inland areas. For example, Portobello compared to Peebles.

Air masses

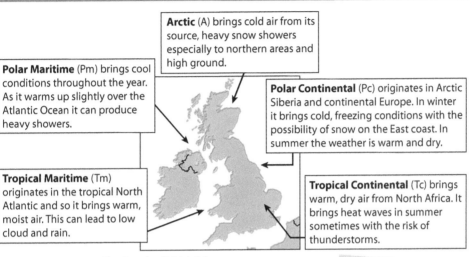

Arctic (A) brings cold air from its source, heavy snow showers especially to northern areas and high ground.

Polar Maritime (Pm) brings cool conditions throughout the year. As it warms up slightly over the Atlantic Ocean it can produce heavy showers.

Polar Continental (Pc) originates in Arctic Siberia and continental Europe. In winter it brings cold, freezing conditions with the possibility of snow on the East coast. In summer the weather is warm and dry.

Tropical Maritime (Tm) originates in the tropical North Atlantic and so it brings warm, moist air. This can lead to low cloud and rain.

Tropical Continental (Tc) brings warm, dry air from North Africa. It brings heat waves in summer sometimes with the risk of thunderstorms.

Fig 1.13 Air masses affecting the British Isles

The weather in the UK is directly affected by the winds that blow over it. These winds originate from large volumes of air called air masses. The characteristics of an air mass depend on where it has spent the last few days or weeks.

- If it has been over the Equator (tropical) it is warm.
- If it has been over the Poles (polar) it is cold.
- If it has been over the sea (maritime) it is wet.
- If it has been over the land (continental) it is dry.

Air streams are movements of air from its source region. The UK lies at the intersection of five air masses, which gives us our changeable weather.

EXAM TIP

Remember to link your knowledge of these weather factors with the reasons behind the presence of different land uses in the case study areas you have already covered in this section, such as the Cairngorms or the Yorkshire Dales.

Quick Test

1. Describe how you could identify from a weather map that the winds associated with a low pressure system or depression are stronger than those in a high pressure system or anticyclone.

2. List the four factors affecting weather in the UK.

3. On a sketch map of the UK, practice drawing and summarising the main characteristics of the air streams affecting it.

4. Why are places warmer at the Equator than at the Poles?

River landscapes

In this section of the course you will be expected to know about rivers and their valleys. You should be prepared to draw diagrams to help you explain how river features such as waterfalls and meanders are formed. You will also be expected to know and be able to identify the river stages shown on the reference maps below.

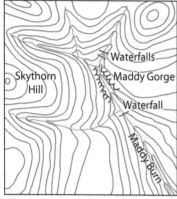

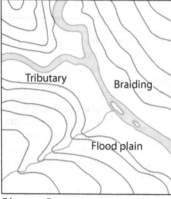

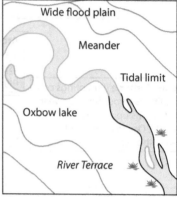

Diagram A
Upper Course Youthful Stage
Waterfalls, gorge, rapids,
interlocking spurs

Diagram B
Middle Course Mature Stage
Meanders, river terraces, braiding
flood plain

Diagram C
Lower Course Old Age Stage
Large meanders, oxbows,
deposition, marshland tidal waters

Fig 1.14 Landscapes on OS maps

The Ordnance Survey type maps (Fig 1.14) show the three typical river stages:

* Near the source in the mountains (Diagram A) the river is in the youthful stage.
* Lower downstream (Diagram B) is the mature stage in the middle of the river's course.
* At the end of the river's journey to the sea (Diagram C) the river goes into its old age stage.

Use the OS style maps and the table below to help you identify and understand the meaning of river features and how they occur on maps.

River stage	River course	River landscape	River features
A. Youthful	Upper course	Mountain stage	V-shaped valleys, waterfalls
B. Mature	Middle course	Valley stage	Meanders, braiding
C. Old age	Lower course	Plain stage	Oxbow lake, levee, flood plain

River processes

Rivers involve three processes: **erosion** of the landscape, **transportation** of sediment and **deposition** of sediment.

Erosion by rivers breaks the landscape down and in National 5 Geography you should be able to explain how rivers erode their valleys using some of the following processes:

Hydraulic action

The river pushes air and water into cracks in the river bed and bank, and the rock breaks up.

Corrasion

Large and small fragments of rock such as pebbles and boulders rub the surface of the river bed, grinding it away.

Attrition

Rock fragments are carried by the river and collide and break up into smaller and smaller fragments.

Corrosion and solution

The river water dissolves the bedrock, for example acid rainwater will dissolve limestone.

Rivers **transport** the eroded materials along and may deposit them later (see meanders and levees on page 21). Rivers move sediment in a number of ways. Large boulders are rolled along by the flow (**traction**), smaller rocks and boulders are bounced along the river bed (**saltation**), fine sediment floats in the flow of the river (**suspension**) and some material is dissolved and carried in the water itself (**solution**).

Quick Test

1. How do rivers erode the landscape?
2. What is hydraulic action?
3. What do the terms corrasion, attrition and solution mean?

The formation of river features

You should be able to recognise river landscape features on diagrams like those on page 18. You will also be expected to explain how some of these features were formed: most typically either a waterfall, a meander or an oxbow lake.

Formation of waterfalls

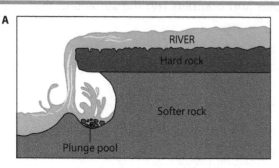

A1
The hard rock protects the softer rock underneath from erosion, but the winter fall undercuts the hard rock.

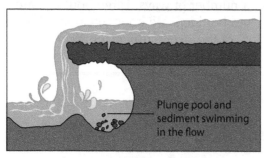

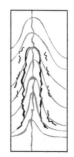

B1
The softer rock is more easily eroded and the hard rock above it becomes increasingly undermined and unsupported. The debris in the pool erodes the softer rock.

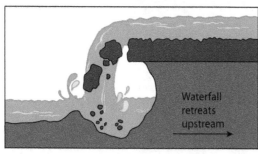

C1
The unsupported hard rock collapses into the plunge pool and the waterfall retreats upstream. This creates the downstream Gorge shown A1, B1, C1

Fig 1.15 Formation of a waterfall and gorge

Waterfalls can be formed when a river flows over alternating beds of hard and soft rock (**A**). Erosion takes place using **hydraulic action**, where air and water is forced into the cracks in the bedrock, or **corrasion**, where the river grinds away at the bedrock using its load of sand and gravel. This erodes the weaker rock more than the harder rock above (**B**). Erosion undermines the hard rock on top until it cannot be supported and collapses into the stream (**C**). In this way, the river cuts backwards into the landscape, forming a waterfall and steep-sided gorge. The maps linked to each part of the diagram (A1, B1 and C1) shows how the steep-sided rocky gorge gets longer and longer as the waterfall retreats.

How do meanders and ox bows form?

Rivers do not flow in a straight line. Contact between the river's water and the sides and bed make the water turbulent. Turbulent flow causes the water in the river to flow from side to side. This is called the river **meandering**. The speed of flow is greater in a meander on the outside of the bend so erosion takes place here (see Fig 1.16). Erosion takes place using **hydraulic action**, where air and water is forced into the cracks in the bedrock, or **corrasion**, where the river grinds away at the bedrock. On the inside of the bend, the flow rate is slower so the river deposits here. Over time, erosion of the outside bends into the neck of the **meander** sees the space in between eroded away more and more, until the river breaks through to form a new course. The old river course is abandoned and an **oxbow lake** is left behind.

> **EXAM TIP**
>
> Keywords for river landscape answers are: hydraulic action, abrasion, attrition, corrosion, undercutting, plunge pool, meander, oxbow lake and deposition.

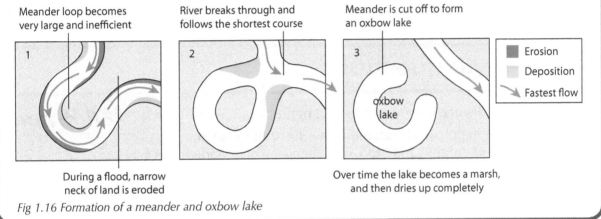

Fig 1.16 Formation of a meander and oxbow lake

Other river processes

Use the erosion processes above to explain the formation of V-shaped river valleys. Braiding takes place where the river current is slow and sand and gravel is deposited by the river which causes the river to break up into several channels called braids. Use the transportation and erosion processes to explain this feature. Levees are found in the lower course where the river has created a low-lying flood plain. As the river bursts its banks its flow slows down as it leaves its channel. This leaves a deposit of sediment which over time builds up a raised bank alongside the river called a levee. Levees may prevent the river from flooding if people build them higher.

Quick Test

1. How does the river undermine the waterfall?
2. What is an oxbow lake?
3. How are levees formed?
4. **Exam-style question**: With the aid of an annotated diagram, explain how waterfalls are formed.
5. **Exam-style question**: Explain the formation of a meander. You may use diagrams to illustrate your answer.

Rivers on Ordnance Survey (OS) maps

You should be able to identify river features on an OS map and explain how they were formed. You also need to describe a river and its valley on an OS map. The diagrams below will help you spot the patterns associated with some of the main river features.

V-shaped river valley

Meander and oxbow lake

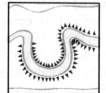
Levees to prevent flooding

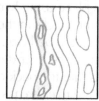
Braided stream

Waterfall and gorge

Fig 1.17 Common river features

Exam-style questions

1. Study the Ordnance Survey map extract opposite.

    a) Describe the physical features of the river Allt Creag an Leith-choin from 972032 to where it enters Loch Morlich at 970093. **(4 marks)**

    b) Identify the features of glacial erosion found at each of the following grid references: 964026, 981028, 988035 and 004026. Choose from corrie and lochan; arête; hanging valley and U-shaped valley. **(3 marks)**

> **EXAM TIP**
>
> In the exam you will answer a question on EITHER rivers and valleys OR limestone. You will only be able to answer Q1(b), Q2(a), Q2(b), Q3(a) and Q3(b) if you have studied Glaciation.

2. a) Using map evidence, describe and explain the advantages and environmental impact of tourism in the area as shown in the Cairngorms map extract. **(3 marks)**

    b) Explain how a corrie is formed. You may use diagram(s) to illustrate your answer. **(4 marks)**

3. Study the Ordnance Survey map extract opposite and the transect diagram below.

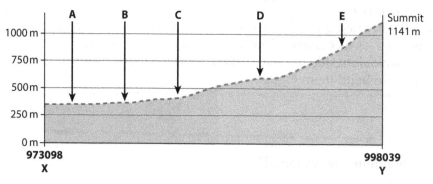

    a) Match the letters shown on the transect X-Y above to the following land features and land uses: ski area, river, forest edge, Fiacaill a' Coire Cais, forest track. **(3 marks)**

    b) To what extent do you agree that ski development may be at conflict with other land uses in the area of the map extract? **(4 marks)**

4. Answer the following question using map evidence.

   a) Describe the leisure and tourism features available on the map. **(3 marks)**

   b) Explain the benefits these leisure and tourism features bring to the area. **(4 marks)**

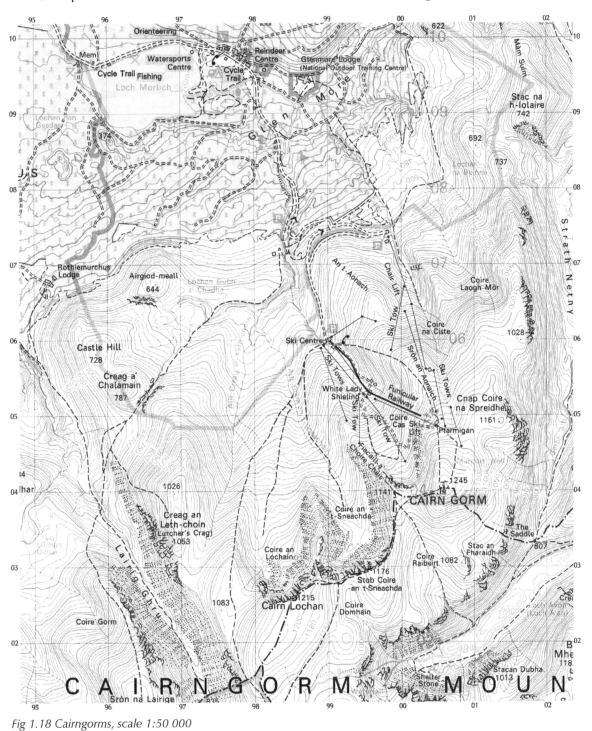

Fig 1.18 Cairngorms, scale 1:50 000

Fieldwork activities

A local river can be used to study river features in the real world, and this is a very effective way of learning about the way that rivers work in the landscape. There are a number of straightforward techniques that you can employ to study a river near you.

Investigating a burn or small stream

To prepare for your visit, follow these steps.

1. Identify the local study area on OS maps, e.g. using Digimaps and on a satellite picture (for example, use Google Earth).

2. Search out a local geology map perhaps from your local library to see if geological features play a part in creating river features.

3. Access any information you can about the river on the Internet.

4. Create a sketch map to scale of the section or contrasting sections of the river under investigation.

5. Draw cross sections of the river at various points along its course, with lines showing the sections on the sketch map, and label the cross sections A–B, C–D etc.

6. Draw a long profile (long section) of the river.

On your field visit, use a digital camera to photograph river features such as waterfalls and meanders, and record the location of each feature on the sketch map. Remember to look at both features of erosion and deposition.

*   Draw annotated field sketches of points of interest you see on your visit to the burn. Include any man-made features, such as flood banks and weirs, to show how the river is being managed and to help judge how these features may change the way the river operates.

*   Use a flow meter to examine the flow across a meander at regular depths and create an Isovel diagram. Use the flow meter to measure flow velocity at various points of the river. Relate these measurements to things such as the stage the river is at, i.e. upper course or middle course etc. Relate these measurements to the cross sections and long profile you have drawn.

*   Measure the velocity of the surface flow over a set distance using a floating object such as an orange or tennis ball, by timing the object from point-to-point several times and taking an average velocity.

Writing up your results

One of the best ways to write up your results is to create a graphic display or an analysis sheet full of maps diagrams, photos and explanatory text. Here is an example.

Study of a local river – the Reekie Linn

The Reekie Linn is a waterfall and gorge on the River Isla in Angus, located where the river leaves the Grampian Highlands and enters the broad valley of Strathmore, near the small town of Alyth. Using the OS map of the area and Google Earth alongside a field visit, it is possible to complete a short study of the river and its spectacular features. In doing so, we can investigate why the waterfall formed here and what created the spectacular gorge.

Fig 1.19 Reekie Linn waterfall

Field study notes

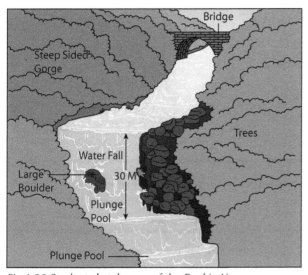
Fig 1.20 Student sketch map of the Reekie Linn

Investigation of the Reekie Linn showed that the river flowed over hard volcanic rock (dolerite) at the point where it crossed the Highland Boundary Fault, which was underlain by softer old red sandstone rocks on the southern side of the fault.

EXAM TIP

Use a field study notebook to hold maps and pre-prepared diagrams and to record notes and sketches taken in the field. It can also store annotated photographs and will in addition provide evidence of work you have done in the field for assessment purposes. A field study notebook will also be useful for helping you to write up an individual study that you will complete in class.

Writing up

The field study notes can be used alongside your knowledge of how waterfalls form to explain how the Reekie Linn (or any other river feature near you) has formed. Measurements of small burns and rivers give you data that can be turned into useful graphs, for example to compare the speed of flow of the river at different stages in its course. Measurements could be taken across a meander to study the different depths and velocity of flow in the feature. Knowledge of local geology might help explain how rock type leads to waterfall formation in some cases. Rivers may be misfits if they have had glacial melt water flowing through them in the past and this could also form the basis of a study. The key thing is to be able to explain the river features you find.

EXAM TIP

You will also need to know the land uses appropriate to rivers and their valleys, including:
- farming
- forestry
- industry
- recreation and tourism
- water storage and supply
- renewable energy

Limestone scenery

In this section of the course you will be expected to know about **limestone scenery**, also called **karst scenery**. You should also be prepared to draw diagrams to help you explain how a range of limestone features are formed. You will also be expected to know and be able to identify on an OS map the range of limestone features detailed in this chapter.

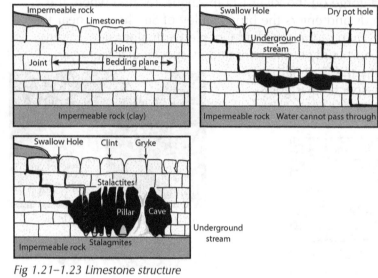

EXAM TIP

- Stalactite – **c** for ceiling, Stalagmite – **g** for ground.
- Stalactites hold on tight.

Fig 1.21–1.23 Limestone structure

How does limestone form?

Limestone is a common rock found in many parts of the UK. It was created from the bodies of sea life in ancient seas. As the creatures died, they formed layers of limey ooze on the sea bed, creating layer after layer of rock beds separated by **horizontal bedding planes**. After the limestone formed it cracked vertically, caused by folding and earth movements and also the limestone shrinking as it dried out. These cracks are called **vertical joints**.

What makes limestone – and in particular carboniferous limestone – special, is the fact that it is soluble in rainwater and this produces a whole range of features found in limestone landscapes.

Diagrams like the one above can be drawn and labelled to help explain the formation of a number of limestone features from above and below ground. You will also have studied an area of limestone scenery, such as the Yorkshire Dales or the Peak District National Park, and this should provide examples of limestone features to add to your answer.

You need to be able to recognise a range of limestone features on a diagram. You should be able to spot these features on model landscape diagrams as well as on OS maps. We will look at some of these features on page 28.

- **Above ground (surface)** – clints, grykes, limestone pavement, swallow hole, pot hole, sink hole, disappearing stream, gorge.
- **Below ground (subterranean)** – cavern, stalagmite, stalactite, pillar.

OS mapwork skills

You should be able to spot the following limestone features on OS maps like the one in this section and on the diagrams provided: limestone pavement, stalagmites and stalactites, limestone gorge, swallow holes and potholes, cave or cavern, clint, gryke, scree slopes, intermittent drainage.

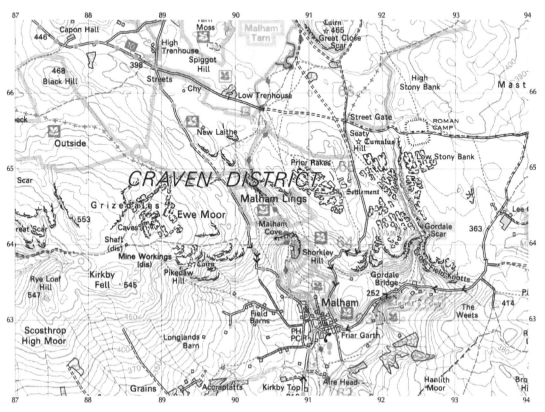

Fig 1.24 Malham, scale 1:50 000

Exam-style question

Study the Yorkshire Dales National Park Ordnance Survey Map extract which shows a variety of limestone features including swallow holes, caves, a gorge and a limestone pavement. Select one of the features from the list and explain how it was formed. You may use diagrams if you wish. **(4 marks)**

Quick Test

1. Why does rain water weather limestone?
2. How do clints, grykes and swallow holes form?
3. What would limestone pavement look like on an O.S. map?
4. How are stalactites and stalagmites created?

Limestone features: exam-style question

Question: How do limestone pavement and other limestone features form?

Answer: Limestone pavement

Limestone is made up of the shells and dead bodies of sea life. (1) It was laid down in beds. Each bed of limestone is separated from others by horizontal bedding planes. (1) As the limestone mass dried out, or as a consequence of earth movements, vertical cracks formed called joints. (1)

Falling rain dissolves carbon dioxide in the atmosphere, which turns it into weak carbonic acid. (1) This dissolves limestone, especially where the landscape was stripped bare of soil by glaciers that left the soluble rock open to attack. (1) Joints and bedding planes are especially under attack and are enlarged over time. (1)

The surface joints are dissolved to form crevices called grykes with upstanding blocks (clints) of limestone inbetween. (1) Large areas of clints and grykes can be found on the limestone uplands and are called limestone pavements because of their block structure. (1)

Other limestone features

Solution also creates other limestone features. Acid rain water forms streams that open up joints and bedding planes to form swallow holes (1) or water sinks (sink holes), where streams disappear underground, leaving dry valleys on the surface. (1) Joints and bedding planes can be opened up beneath the surface and eroded into massive caverns and connecting passages as water flows underground. (1) Underground streams eventually meet impermeable rock layers beneath the limestone, such as clay, and reappear on the surface as a resurgent stream. (1) These often resurface at the foot of a limestone scarp or scar (steep almost vertical rock face). (1)

Underground features

Stalagmites and stalactites are formed in caves and caverns from the water dripping from the ceiling of the cave. (1) This water is charged with calcium carbonate in solution. As the water drips from the ceiling of the cave, evaporation leaves behind a tiny deposit of calcium carbonate mineral called calcite. (1) This builds up like a stone icicle hanging from the roof of the cave and is called a **stalactite**. (1) The drips fall to the ground below, where calcite is deposited again to form a **stalagmite** (1), which grows upwards towards the roof. If a stalactite and stalagmite meet and join together, they form a pillar. (1)

EXAM TIPS

- Here are some keywords that will help with answers to limestone questions: joints, carbonic acid, permeable, impermeable, soluble, calcium carbonate, calcite.
- If the question says 'you may use a diagram(s)' then **think strongly about** using a diagram(s) to help your explanation. Diagrams score marks!
- Learn the landscape features you need in your answers by studying the diagrams provided here or in your notes. Practise drawing and labelling them.

Land use and land use conflict in limestone areas

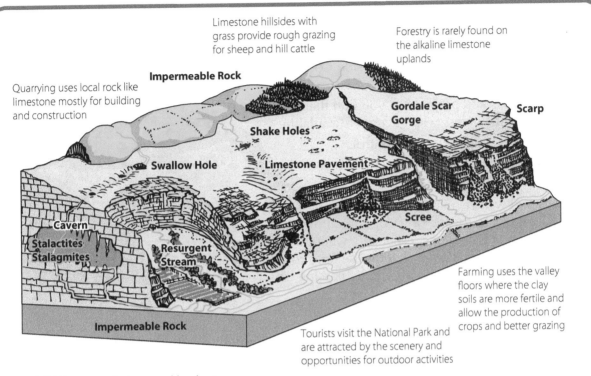

Limestone hillsides with grass provide rough grazing for sheep and hill cattle

Forestry is rarely found on the alkaline limestone uplands

Quarrying uses local rock like limestone mostly for building and construction

Impermeable Rock

Gordale Scar Gorge

Scarp

Shake Holes

Swallow Hole

Limestone Pavement

Cavern
Stalactites
Stalagmites

Resurgent Stream

Scree

Impermeable Rock

Farming uses the valley floors where the clay soils are more fertile and allow the production of crops and better grazing

Tourists visit the National Park and are attracted by the scenery and opportunities for outdoor activities

Fig 1.25 Limestone features and land use

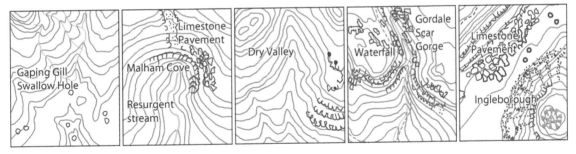

Gaping Gill Swallow Hole

Limestone Pavement

Malham Cove

Resurgent stream

Dry Valley

Waterfall

Gordale Scar Gorge

Limestone Pavement

Ingleborough

Fig 1.26 Limestone scenery

Limestone areas like the Yorkshire Dales are dominated by **sheep and hill cattle farming** because of the cool climate and poor soils found on the limestone hills. Crops can only be grown in the sheltered valley bottoms on more fertile clay soils. Farms tend to be large and employ a lot of rough grazing.

Limestone is used widely in the construction industry for building stone and for cement making. It is also used by the chemical industry, so **quarrying** is an important economic activity in areas like the Yorkshire Dales and the Peak District National Parks.

EXAM TIP

Use the diagram from this section to help you to visualise the limestone landscape being talked about.

What land use conflicts occur in limestone areas?

Farming and tourism

Conflicts can break out between farmers and tourists in the National Parks, especially if visitors do not follow relevant parts of the Countryside Code (www.naturalengland.org.uk):

- Leave gates and property as you find them and follow paths unless wider access is available.
- Keep dogs under effective control.
- Leave no trace of your visit and take your litter home.

Tourists and walkers may cause problems if they allow livestock to escape or if dogs are allowed to run loose among livestock, especially where there are lambs or calves. Livestock let loose onto roads is a traffic hazard and can be injured by passing cars.

Conflict with farmers can be resolved in various ways:

Fig 1.27 Dogs should be kept under close control

- Providing information to visitors about behaving responsibly in the countryside. The country code lays out rights and responsibilities, and citizenship education in schools underpins this.
- Farmers help manage visitor pressures by providing good access routes with gates and stiles, and provide notices and signs that guide walkers and visitors using their land.
- Dogs are kept on leads or under close control.
- Fences and walls are well maintained and footpaths are built where erosion has been caused by too many people using a particular route.

Limestone quarrying

Limestone for cement and building material is an important economic resource and quarrying and cement production employs many thousands of people in Britain. It therefore makes a significant contribution to the economy. It is an activity that has gone on for generations, even in the National Parks.

Quarrying does have environmental effects and can come into conflict with the needs of those who wish to conduct tourist activities. People object to quarrying and cement production for a variety of reasons:

- Excavation and processing limestone creates ugly scars in the landscape, spoiling local views.
- Blasting operations cause disturbance and can close off nearby areas.
- Large amounts of dust are generated and this can affect the landscape nearby.
- Narrow country roads are affected by heavy lorries carrying rock and cement, causing congestion and traffic hold-ups at busy times.
- Wildlife may be disturbed.

Against this, the quarry companies point out that the industry plays an important part in the economy and was always present in the National Park areas. They also employ many local people in areas that have few other jobs.

Conflict with the quarrying industry is resolved by agreements that manage the impact of excavation of limestone and its manufacture into cement.

- Screens of trees help mask or reduce the visual impact of quarries.
- Blasting may be restricted to times that cause the least disturbance to tourism and to local residents.
- HGV movement on the roads is scheduled to avoid busy times.
- Quarries are damped down with water to reduce dust and HGV trailers are covered to reduce dust dispersal.
- Railways and special wagons are used instead of road transport.
- Old quarry workings are restored to good condition or are converted to nature friendly sites by planting woodland, or establishing ponds and wetlands suitable for water birds and other wildlife.
- Derelict buildings are removed.
- Rock faces can be used for climbing, helping the local economy once quarrying has stopped.

Quick Test

1. Why might quarrying disturb wildlife?
2. How might quarrying affect tourist income?
3. How effective are measures taken to control the impact of quarrying?
4. **Exam-style question:** For an area you have studied, describe one example of land use conflict and explain how it was resolved. Use the information on Fig 1.25 to help you.

Glaciation

The starting point of a glacier is a hollow high on the mountainside, a corrie. Snow accumulates above the snowline. Below the snowline is the ablation zone where the snow melts in summer and the glacier begins to retreat.

Fig 1.28 Mer de Glace Glacier, France

EXAM TIPS

If you have not studied the rivers and limestone topics, you will have covered glaciation and coasts for your choice question in section 1 in the exam question paper.

Key processes

- **Erosion**– As a glacier moves downhill it wears away the land it moves over. (See plucking and abrasion below.)
- **Transportation**– The movement of material by the glacier. Referred to as load or moraine.
- **Deposition** – At the snout (the end of the glacier), the ice melts and its load is dumped or deposited.

EXAM TIP

Use the key processes when explaining landform formations.

Processes of erosion

Ice is a powerful agent of erosion in two ways:

1. **Plucking** – ice freezes onto rock and when the ice moves it pulls a piece of rock away: the tweezer effect.

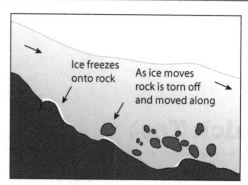

Ice freezes onto rock

As ice moves rock is torn off and moved along

Fig 1.29 Plucking

2. **Abrasion** – pieces of rock carried by the ice scrape at the surrounding rock and erode it away: the sandpaper effect.

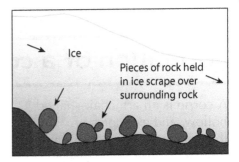

Fig 1.30 Abrasion

Freeze-thaw weathering is also wearing away the landscape. Water in small cracks in rock will freeze, expand and widen the cracks. Repeated freezing and thawing will cause rocks to shatter. A scree slope can form.

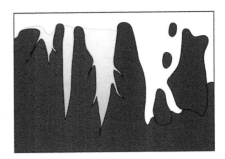

Fig 1.31 Freeze-thaw weathering

Named glaciated uplands

Some of the UK's glaciated uplands are now designated as National Parks. Make sure you can locate upland glaciated areas on a map of the British Isles.

Key:

1: Cairngorms

2: Loch Lomond and The Trossachs

3: The Lake District

4: Snowdonia

Fig 1.32 Map of UK National Parks

Quick Test

1. What are the key processes of glacial erosion?

2. What work does a glacier do?

Landforms of erosion

Formation of a corrie

A corrie is an armchair-shaped hollow, and examples of this can be seen at Coire an t-Sneacdha or Coire Cas.

Fig 1.33 A corrie

Before glaciation	During glaciation	After glaciation

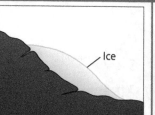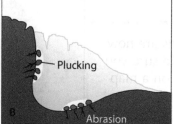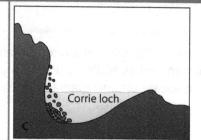

Fig 1.34 Formation of a corrie

• Snow accumulates in a hollow (usually north-facing). • Snow is compressed into neve, then glacial ice. • Freeze-thaw action takes place at the top of the hill where rock is exposed.	• The glacier moves downhill because of gravity, in a rotational manner. • Plucking steepens the back wall. • Abrasion deepens the hollow.	• The ice melts, leaving a hollow in the mountain side. • A rock lip may form, caused by rotation. • Melt water may be trapped by moraine and form a corrie lochan or tarn. • A scree slope may form.

When two or more corries are found back-to-back and a steep knife-edge ridge is formed it is called an **arête**, such as Helvellyn in the Lake District. When three or more corries form back-to-back a frost-shattered, horned peak called a **pyramidal peak** can occur, for example the Matterhorn in Switzerland.

Three-stage formation of a U-shaped valley

Fig 1.35 Glen/Loch Avon

> **EXAM TIP**
>
> Make sure you have named examples of key features to enhance your answer.

Before glaciation	During glaciation	After glaciation
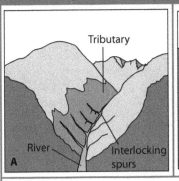	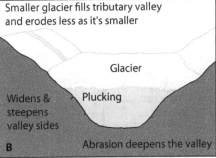	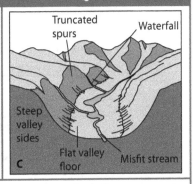

Fig 1.36 Formation of a U-shaped valley

• A river flows in a V-shaped valley around interlocking spurs	• A glacier fills the river valley, moving downhill because of gravity, following the path of the existing river. • The weight/volume of ice combined with plucking, abrasion and rotational sliding will deepen, widen and steepen the valley (bulldozer).	• The ice melts to leave a valley with a flat floor and steep sides: U-shaped. • Interlocking spurs are now truncated. • A misfit stream – too small to have eroded the valley it flows in. • Hanging valley – a smaller valley created by a smaller tributary valley, suspended above the main valley.

Quick Test

1. What are the main processes of erosion and how do they work?
2. Name some of the main glacial landscape features in an area you have studied.

Case study – Cairngorms National Park

Cairngorms National Park, home to Britain's highest and largest mountain range, was established in 2003. It covers an area of 4,528km². It is a good example of a glaciated upland that provides a wealth of land uses. It is managed and protected by a number of organisations including the National Park Authority. See http://cairngorms.co.uk/look-after for further information.

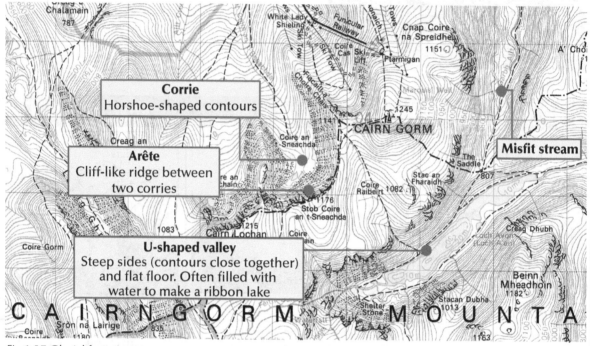

Corrie
Horshoe-shaped contours

Arête
Cliff-like ridge between two corries

Misfit stream

U-shaped valley
Steep sides (contours close together) and flat floor. Often filled with water to make a ribbon lake

Fig 1.37 Glacial features

Land uses

Tourism: in summer, or more importantly during the main winter ski season, the Cairngorms is visited by up to 5000 tourists a day. Visitors use the funicular railway to access the mountain for skiing, or to visit the Ptarmigan Restaurant. The area is popular for walking and mountain biking or sightseeing (passive tourism).

EXAM TIP

You will need to be able to identify and describe glacial features on an OS map including u-shaped valleys, corries, aretes, truncated spurs and pyramidal peaks.

Fig 1.38 Glenshee Ski Centre, Cairngorms

Conservation: the National Park is protected by law but many conservation groups take an interest in preserving the natural landscape and managing its use.

Farming: the steep valley sides and thin soils are ideal for sheep grazing or keeping deer to be sold for venison. In the flat valley floors, some crops can be grown, such as barley, or hay for animal feed.

Quarrying: granite is quarried for building purposes, making good use of the natural stone.

Forestry: plantation forestry is ideal for growing in the sub-arctic climate. The Caledonian Scots Pine is being re-introduced as a native species.

Fig 1.39 Deer, Cairngorms National Park

EXAM TIP

You must consider how a glaciated area such as the Cairngorms has a wide variety of land uses, including:
- farming
- forestry
- industry
- recreation and tourism
- water storage and supply
- renewable energy

EXAM TIP

Land users can be in conflict with each other – you must know the conflict and solutions to the conflict, for a named case study area.

Quick Test

1. Explain how a corrie is formed.
2. What conflicts might arise in a National Park between tourists and other users?
3. How can National Parks be managed?

The coast – erosion

The coast is the boundary between the land and sea.

The coastal system

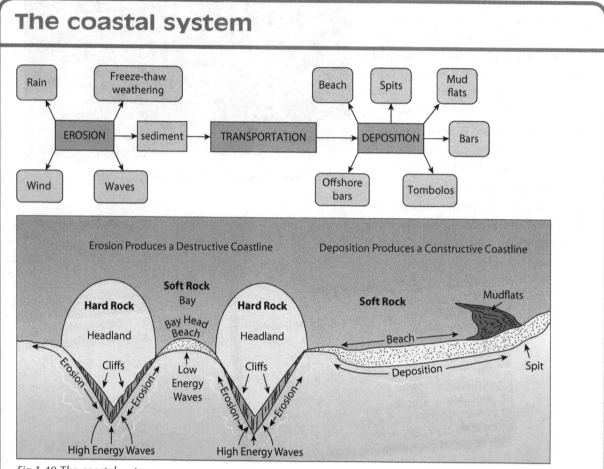

Fig 1.40 The coastal system

Key processes

There are three key coastal processes:

- **Erosion** – as the waves move, they wear away the coastline. See hydraulic action, attrition, corrosion and corrasion below.
- **Transportation** – the movement of material by the wave. See constructive and destructive waves and longshore drift on page 42.
- **Deposition** – when the waves run out of energy, material is dumped.

Erosion: what factors determine the shape of the coast?

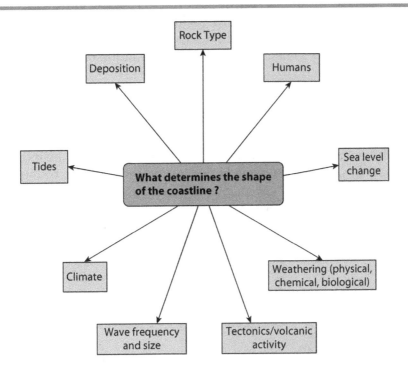

Waves

Waves are the most effective agent of erosion along the coastline. They are created by the wind blowing over the surface of the sea. Their size is determined by:

*   the fetch – the distance over which the wave travels
*   the wind speed
*   the depth of the water – this controls wave height

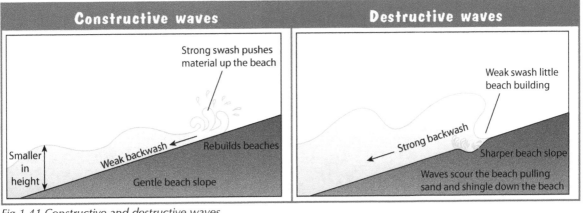

Fig 1.41 Constructive and destructive waves

Coastal landforms resulting from erosion

The sea erodes in four ways:

- **Hydraulic action** – the power of the waves hitting the cliffs causes water and air to be forced into the cracks in the rocks at high pressure. This causes the cracks to widen, splitting a section of cliff. When the waves cut away at the base of a cliff, the process is called undercutting.
- **Attrition** – The materials being carried by the sea are being constantly worn down by knocking into each other, making them into smaller, rounder particles.
- **Corrosion** – Softer rock such as chalk or limestone is gradually dissolved by sea water.
- **Corrasion** – This is the wearing away of the coastline by the erosive effect of sand, shingle and boulders, i.e. a load being hurled against the coastline by the waves.

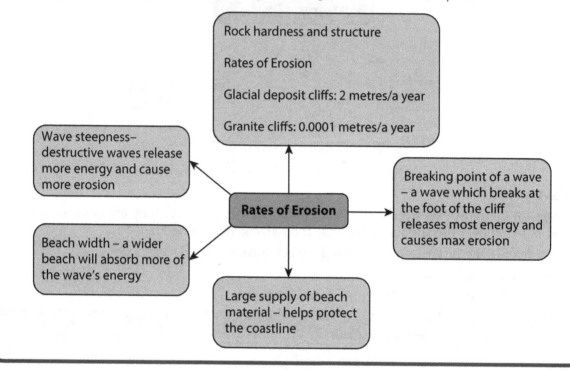

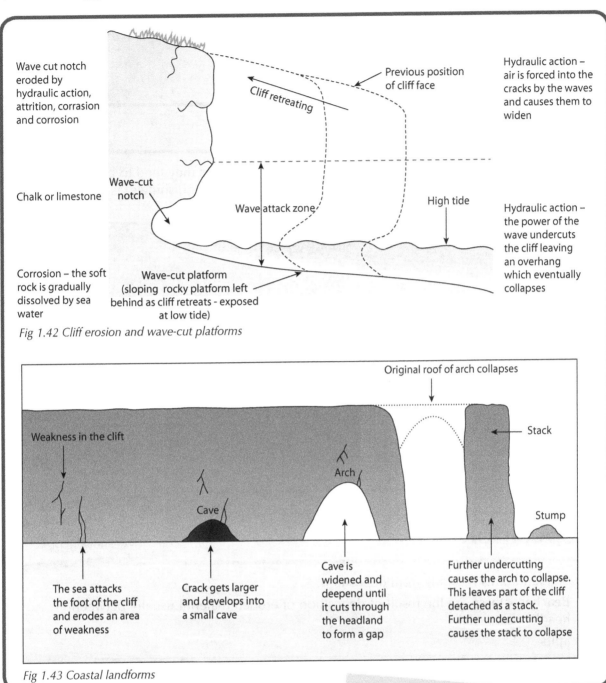

Fig 1.42 Cliff erosion and wave-cut platforms

Wave cut notch eroded by hydraulic action, attrition, corrasion and corrosion

Cliff retreating

Previous position of cliff face

Hydraulic action – air is forced into the cracks by the waves and causes them to widen

Chalk or limestone

Wave-cut notch

Wave attack zone

High tide

Hydraulic action – the power of the wave undercuts the cliff leaving an overhang which eventually collapses

Corrosion – the soft rock is gradually dissolved by sea water

Wave-cut platform (sloping rocky platform left behind as cliff retreats - exposed at low tide)

Fig 1.43 Coastal landforms

Original roof of arch collapses

Weakness in the clift

Stack

Arch

Cave

Stump

The sea attacks the foot of the cliff and erodes an area of weakness

Crack gets larger and develops into a small cave

Cave is widened and deepend until it cuts through the headland to form a gap

Further undercutting causes the arch to collapse. This leaves part of the cliff detached as a stack. Further undercutting causes the stack to collapse

EXAM TIP

You also need to know how headlands and bays are formed, and the importance of differential erosion in their formation.

Quick Test

1. What 3 factors determine the size of a wave?
2. Identify two differences between a constructive wave and a destructive wave.
3. What happens in the process of undercutting?

The coast – transportation and deposition

Transportation

Remember, waves rarely approach the coast at right angles. As they tend to approach it at an angle determined by the wind direction, eroded material (sand/shingle) is moved in a zigzag course along the coast.

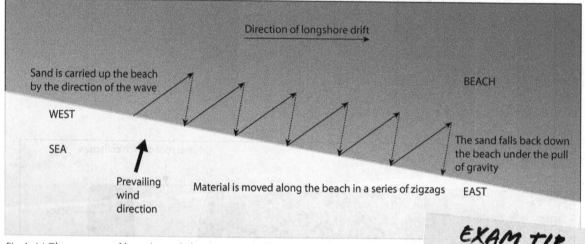

Fig 1.44 The process of longshore drift

Deposition

Coastal landforms resulting from deposition:

- **Beaches** – these are the result of deposition of eroded material usually between headlands.
- **Spits**

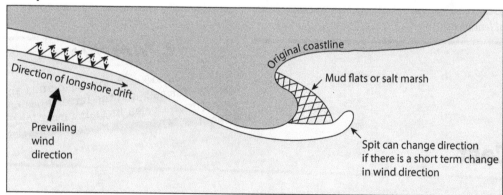

Fig 1.45 Spit

- These are long, narrow ridges of sand or shingle.
- They occur where the coastline changes direction.
- They extend into the sea or part-way across a river estuary.
- Sand spits are formed by constructive waves.
- Shingle spits are formed by destructive waves.
- Many spits develop a hooked or curved end due to a change in the prevailing wind direction.
- Examples of spits: Spurn Head, the Holderness coast.
- **Bars –** if a spit extends across a bay that links two headlands, or across a river estuary, it is called a bar. An example of a bar is Slapton Sands, South Devon.

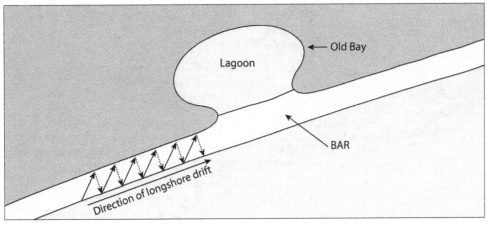

Fig 1.46 Bar

- **Barrier beach –** this is where a spit extends across a shallow bay. An example of a barrier beach is Slapton Sands, south Devon.
- **Tombolo –** this occurs when a sand spit grows outwards into open water and reaches an island. An example of a trombolo is Chesil Beach in Dorset (this joins the Dorset coast with the Isle of Portland).
- **Sand dunes –** dry sand is picked up off the beach by the wind and then it gets trapped by seaweed or driftwood. It gradually piles up until it is colonised by vegetation and in time grows in size to form dunes. An example of a sand dune is Gullane, East Lothian.

Quick Test

1. What is meant by the 'fetch' of a wave?
2. How do constructive and destructive waves differ?
3. List four ways in which the sea erodes and name three examples of landforms resulting from coastal erosion.
4. Describe the process of longshore drift.

Coastal land uses

Very little of the coastline of the UK is 'unused' – the vast majority of our approximately 5,000-mile coast has a wide variety of land uses (or activities) and habitats, both human and natural.

You will have studied particular case studies but you will soon find that there are land uses and conflicts/themes that are common features of many coastal areas. Take the famous Dorset coastline in southern England:

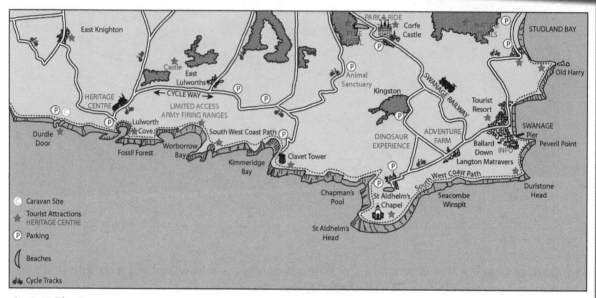

Fig 1.47 The Dorset coast

Pressures on the coastline

- **Retreating or crumbling cliffs** – caused by increased erosion, quarrying, caravan sites, extreme weather.
- **Pollution** – rubbish dumped or washed up on beaches: particularly plastics, sewage. UK beaches are under pressure to meet EU standards of cleanliness (Blue Flag beaches).

Fig 1.48 Waste on a Scottish beach

- **Industrial development** – natural harbours and their surrounding flat hinterlands attract shipping and industry, for example Milford Haven, which is a deep water port, South Wales, and Grangemouth oil refinery, West Lothian.

- **Military uses** – these have restricted local and tourist access due to live firing, for example Tyneham and Worbarrow Bay, South Dorset, and the Kirkcudbright coast, South West Scotland.

- **Increased tourism** – disadvantages: increase traffic, pollution, noise, litter, increased house prices, second homes; advantages: jobs, increased local income.

- **Reduced supply of sediment to beaches, dune systems, spits and bars** – this is caused by hard engineering further up the coast, for example harbours, sea walls, groynes, marinas, jetties and piers may interrupt longshore drift.

- **Global warming** – When sea levels rise there is a greater incidence of extreme weather events such as storms and hurricanes.

Fig 1.49 Grangemouth oil refinery

EXAM TIP

In the context coasts you need to know:
- the conflicts which can arise between land uses within this landscape
- the solutions adopted to deal with the identified land use conflicts

Fig 1.50 Crowded beach, Weston-Super-Mare

Quick Test

1. Study the map and identify all the land uses along this stretch of coast.

2. What pressures might there be on this coastline?

Population distribution

Population distribution is all about **where people live and why**.

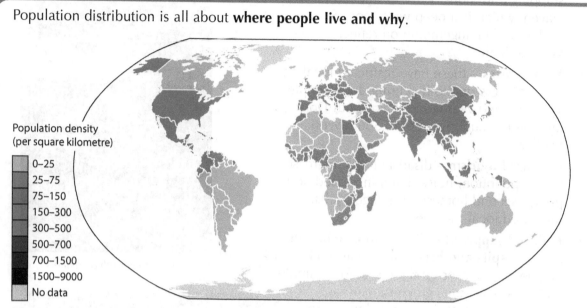

Population density (per square kilometre)

- 0–25
- 25–75
- 75–150
- 150–300
- 300–500
- 500–700
- 700–1500
- 1500–9000
- No data

Fig 2.1 Worldwide population distribution

Population is unevenly distributed around the world. Some places are crowded and have a high population density and other areas are empty and have a low population density.

Population density is a measurement of people per square kilometre. It is calculated by the formula: Area(sq km) ÷ Total Population.

Below there is information about a selection of countries of varying population densities including some of the positive and negative reasons why people choose to live in one area of the world and not in another.

UK

Land area: 244,820 km²

Total population: 66,181,585

Average population density (people per km²): 247

Positive characteristics:	Negative characteristics:	Main influencing factors determining population density:
• Moderate climate (i.e. reliable precipitation and sunshine) • Areas of flat, fertile land either in river valleys or lowland plains • Natural resources (e.g. fossil fuels – coal, gas and oil) • Coastal plains suitable for industry and trade	• Areas of high ground with steep slopes and poor soils • Higher than average precipitation and low temperatures, e.g. Scottish Highlands, Snowdonia and Lake District	• Climate • Availability of flat land • Fertile soils • Availability of natural resources • Accessibility (good road and rail links and ports)

Kenya

Land area: 582,650 km²

Total population: 49,699,862

Average population density
(people per km²): 58

Positive characteristics:	Negative characteristics:	Main influencing factors determining population density:
• Fertile highlands • Flat coastal plain	• Tropical (along the coast) to arid climate with the risk of drought • Scarcity of surface water resources • Lack of natural resources • Disease, e.g. malaria, biharzia, schistosomiasis • Little government investment	• Climate • Water scarcity • Lack of natural resources • Mosquitoes and other disease spreading insects

Bangladesh

Land area: 144,000 km²

Total population: 164,669,751

Average population density
(people per km²): 1,002

Positive characteristics:	Negative characteristics:	Main influencing factors determining population density:
• Monsoon climate with reliable precipitation • Plenty of flat land • Fertile river valleys and coastal plains	• Lack of natural resources • High risk of flooding	• Fertile soils • Flat land

Germany

Land area: 357,021 km²

Total population: 82,194,467

Average population density
(people per km²): 231

Positive characteristics:	Negative characteristics:	Main influencing factors determining population density:
• Moderate climate • Fertile and sheltered river valleys, e.g. the Rhine • Plenty of flat land • Good accessibility • Plenty of natural resources, e.g. coal, iron ore in the Ruhr valley	• None	• Climate • Natural resources • Flat land • Accessibility • At the heart of Europe • Large Government investment

The table on the previous page should help you see why people choose one area of the world to live over another area. Key sustainability concepts to consider include:

EXAM TIP

Make sure you can identify a list of reasons for why some places are empty and others are crowded.

- **Optimum population** – where the amount of resources can support the population.
- **Underpopulation** – where there are too few people to utilise the resources available.
- **Overpopulation** – where there are not enough resources to support the population.

The following photographs and accompanying table show how we might assess the positive and negative characteristics of various parts of the world and assess their suitability to sustain a population.

Photo	General description	Positive or negative area	What are the factors?
Fig 2.2 Great Basin Desert, Utah, USA	Arid desert	Negative	• Lack of precipitation • Water scarcity • No soil, bare earth
 Fig 2.3 New York, USA	Large natural harbour	Positive	• Sheltered harbour • Flat land either side of the Hudson River • Access to the Atlantic Ocean and Europe

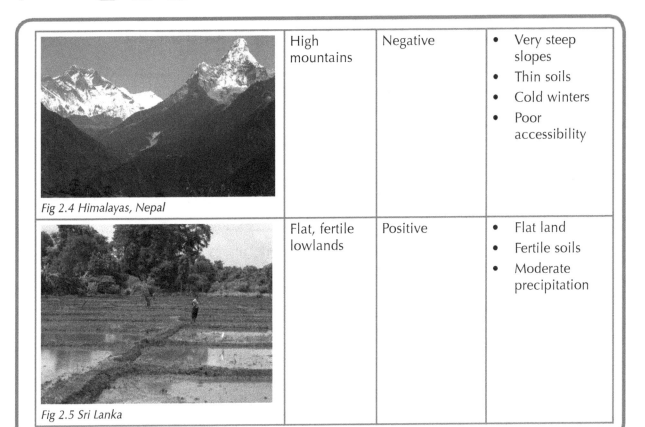

	High mountains	Negative	• Very steep slopes • Thin soils • Cold winters • Poor accessibility
Fig 2.4 Himalayas, Nepal			
	Flat, fertile lowlands	Positive	• Flat land • Fertile soils • Moderate precipitation
Fig 2.5 Sri Lanka			

Quick Test

1. Identify 3 positive characteristics which have encouraged a higher population density.
2. Identify 3 negative characteristics which have lead to a lower population density.
3. Give an example of an area with a higher population density and explain why it has such a high density.
4. Give an example of an area with a low population density and explain why it has such a low density.

World population growth

The world's population has now reached over 7 billion. If you look at the world population graph below you will see that there has been a population explosion (where population grows very rapidly) since 1930. This explosion has largely been the result of a rapid rise of the number of people living in the developing world.

Why does the population change?

The rate at which the world's population increases or decreases depends on the 'balance' between birth rates, death rates and migration.

- **Birth rate** – the number of live births per 1000 people per year.
- **Death rate** – the number of deaths per 1000 people per year.
- **Migration** – the number of people moving in or out of an area.

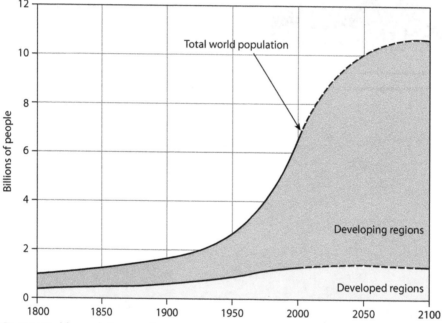

Fig 2.6 World population graph

The difference between the birth rate and the death rate is called natural increase. With a **natural increase**, the birth rate is higher. With a **natural decrease,** the death rate is higher. Since the beginning of the 20th century there has been a high natural increase as the birth rate has exceeded the death rate.

Life expectancy

Life expectancy is the average number of years a person is expected to live. This indicates a good health care system and a well-balanced diet in a country, i.e. Japan's life expectancy is 84 years but South Africa's is only 57 years.

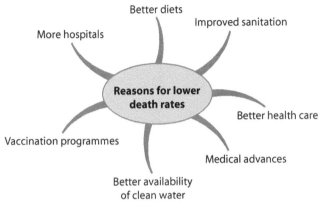

- Better diets
- Improved sanitation
- More hospitals
- **Reasons for lower death rates**
- Better health care
- Vaccination programmes
- Medical advances
- Better availability of clean water

EXAM TIP

Remember – population change can also occur because of natural disasters, e.g. tsunamis and war.

EXAM TIP

Life expectancy is a social indicator of a country's level of development. You should know and understand the use of both social and economic indicators.

Fig 2.7 Japan has one of the highest life expectancies in the world

Quick Test

1. Write down the definitions of birth rate, death rate and natural increase.
2. What other variable can cause population change in an area?
3. Give three reasons for death rates falling worldwide.
4. Give two reasons why developing countries have high birth rates and two reasons why in some developed countries (i.e. Hungary and Germany) the birth rate has recently fallen below the death rate.

Population structures in developed and developing countries

Population structure is shown by a type of graph called a population pyramid. It shows the number of males and females in different age groups.

Developing country – Kenya

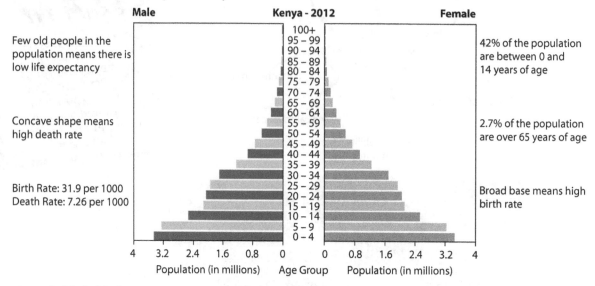

Few old people in the population means there is low life expectancy

Concave shape means high death rate

Birth Rate: 31.9 per 1000
Death Rate: 7.26 per 1000

42% of the population are between 0 and 14 years of age

2.7% of the population are over 65 years of age

Broad base means high birth rate

Kenya's high birth rate

In Kenya there is less birth control, a lack of education and cultural and religious pressures.

Some viewpoints are:

– 'I want to hand my land down to my sons.'

– 'So many children die here before they are five, I am going to have a lot so that I am sure at least some will survive.'

– 'I need my children to help in the fields and make me important in my village.'

Fig 2.8 A large Kenyan family

Developed country – United Kingdom

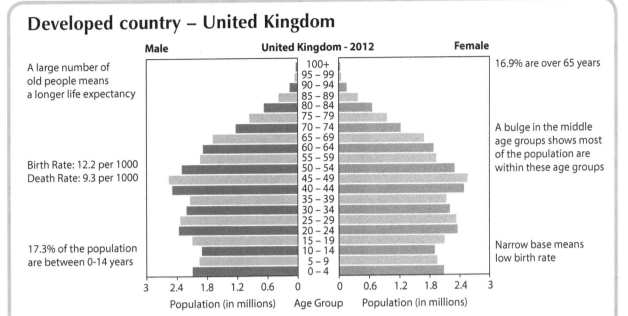

Male   United Kingdom - 2012   **Female**

A large number of old people means a longer life expectancy

16.9% are over 65 years

A bulge in the middle age groups shows most of the population are within these age groups

Birth Rate: 12.2 per 1000
Death Rate: 9.3 per 1000

17.3% of the population are between 0-14 years

Narrow base means low birth rate

Age groups (top to bottom): 100+, 95 – 99, 90 – 94, 85 – 89, 80 – 84, 75 – 79, 70 – 74, 65 – 69, 60 – 64, 55 – 59, 50 – 54, 45 – 49, 40 – 44, 35 – 39, 30 – 34, 25 – 29, 20 – 24, 15 – 19, 10 – 14, 5 – 9, 0 – 4

Population (in millions): 3, 2.4, 1.8, 1.2, 0.6, 0 | 0, 0.6, 1.2, 1.8, 2.4, 3

Population (in millions)   Age Group   Population (in millions)

The UK's low birth rate

In the UK, contraception is readily available. More women work compared to women in Kenya, and family planning is countrywide.

Some viewpoints are:

- 'Children cannot work until they are 16 so it will cost me a great deal to have a large family.'
- 'Contraception is freely available and it is very rare for a child to die.'

Fig 2.9 A British family

Quick Test

1. What is the graph called that shows population structure?
2. If the population pyramid has a broad base what does this indicate about the population structure?
3. If the population pyramid is concave what does this indicate about the population structure?

How does the population change over time?

The Demographic Transition Model is a multiple line graph that shows how population changes over time.

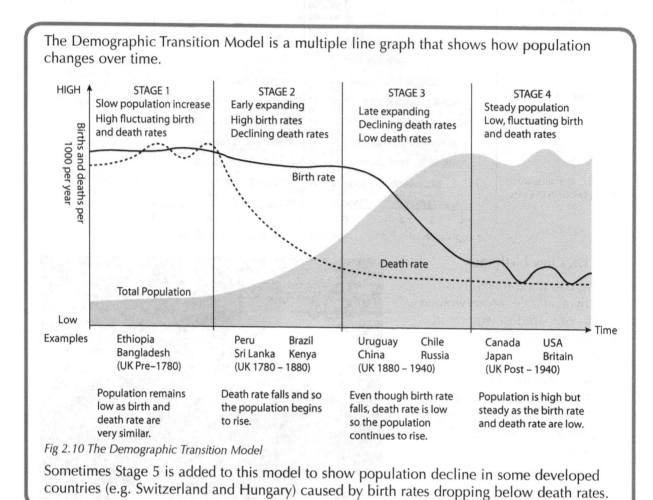

Fig 2.10 The Demographic Transition Model

Sometimes Stage 5 is added to this model to show population decline in some developed countries (e.g. Switzerland and Hungary) caused by birth rates dropping below death rates.

What problems are created by a changing population?

Within any population there is a **dependant (ageing and youthful)** population – these are the people of non-working age, either 0–15 years or over 65 years. They are dependent on the **economically active** population – those people between 16 and 64 years of age.

Developed countries	Developing countries
High number of elderly (**ageing**) dependants because of increased life expectancy.	High number of young (youthful) dependants, because of high birth rates.
Problems caused by this trend: • Older people require more spending on health care • Greater need for public services, e.g. social services, hospitals • Increased number of care/retirement homes required • Increased taxation of the economically active population to fund these requirements • Increased strain on a country's finances • Fewer young people means there will be less of an economically active population in the future	Problems caused by this trend: • Overcrowding • Poverty • Unemployment • Education and health care services are not able to meet needs • Inadequate housing – informal settlement and/or shanty towns spring up • Lack of formal employment leads to the growth of informal sector jobs, e.g. hawking • Increased crime rates • Lack of infrastructure and basic facilities, e.g. clean water, waste disposal

Quick Test

1. What are the four stages of the Demographic Transition model?

2. What is the impact of an ageing population?

Development

Development is the improvement in the standard of living in a country. This can mean many things, such as better education, access to clean water, or improved job opportunities.

Measuring development

We use indicators of development: social, economic, composite. An indicator has to be something that we can measure so that we can compare between one country and another.

Social and economic indicators of development

Social indicators	Economic indicators
• Doctors per 100,000 of the population. • Adult literacy rates. • Number of children attending secondary school. • Life expectancy.	• Average income per person. • Gross domestic product (GDP) – the value of goods and services produced by a country. • Gross national product (GNP) – how much a country earns. Includes value of goods and services involved in trade. It is measured in US$ per person.

Fig 3.25 Indicators of development

Composite indicators of development

We need to use more than one indicator to give an accurate picture so it's best to use composite indicators of development:

- Physical Quality of Life Index (PQLI) – this combines life expectancy, adult literacy rate and infant mortality rate. Each of the indicators is given a rating of between 0 and 100, with 0 = the worst and 100 = the highest.
- Human Development Index (HDI) – a social welfare index. It is a composite of educational attainment (adult literacy and years in education), life expectancy at birth and income per capita (per person).

Indicators of development can be very useful in telling us whether a country is developed or developing.

	UK	France	Ghana	Kenya
Birth rate (live births per 1000 people per year)	11	13	29	38
Death rate (deaths per 1000 people per year)	10	8	9	10
Life expectancy	79	81	59	57
GNP per person (US$)	33 800	33 600	1330	1540
Doctors per 100,000 people	230	336.7	15.2	13.9
Internet usage (% of population)	82	79.6	14.0	28

Fig 3.26 Composite indicators of development

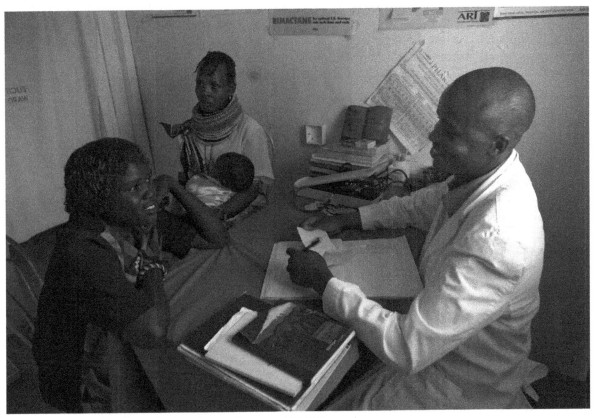

Fig 3.27 In Kenya there are only 13.9 doctors for every 100,000 people

Quick Test

1. Using the table above, which country is most developed and why?
2. What are the problems in using only one or two of the indicators listed?

Urban land use zones

What do we mean by the term urban?

The term **urban** refers to large towns, cities and conurbations. A **conurbation** is a large, **continuous built-up area** created by towns and cities expanding and growing together.

Characteristics of land use zones in the developed world

The term **land use zone** refers to the way that people use areas of a city or town. Often it is useful to think of these in diagrammatical form and to look at a **transect** from the centre of a city to the edge like the one opposite. A transect defines a series of zones that transition from sparse rural farmhouses to the dense urban core.

Urban land use models

In large towns and cities, it is often possible to see a clear pattern of land use which reflects the development of the settlement. The distribution of urban land use can be explained using **models of urban land use** such as those shown below. In Fig 2.12, the concentric rings model is evident.

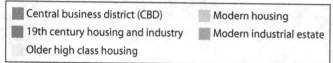

- Central business district (CBD)
- 19th century housing and industry
- Older high class housing
- Modern housing
- Modern industrial estate

Concentric rings model where the city has grown outwards from its old centre.

Sector model where the city has grown around transport lines such as canals or railway lines.

Multiple nuclei model where the city has grown around more than one centre (or CBD).

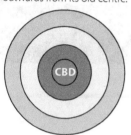

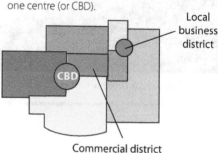

Local business district

Commercial district

Fig 2.11 Urban land use models

Note the contrast between developed world cities and developing world cities (see pages 61–65).

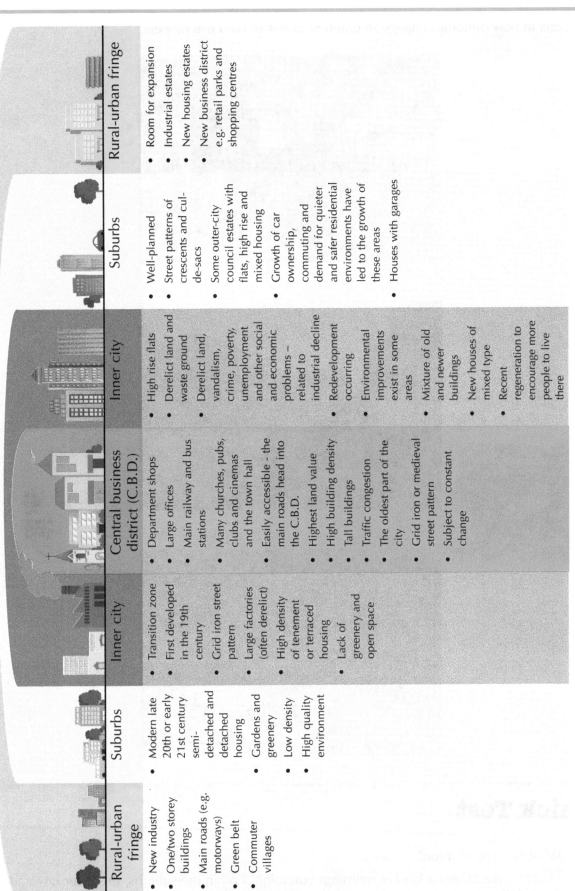

Rural-urban fringe	Suburbs	Inner city	Central business district (C.B.D.)	Inner city	Suburbs	Rural-urban fringe
• New industry • One/two storey buildings • Main roads (e.g. motorways) • Green belt • Commuter villages	• Modern late 20th or early 21st century semi-detached and detached housing • Gardens and greenery • Low density • High quality environment	• Transition zone • First developed in the 19th century • Grid iron street pattern • Large factories (often derelict) • High density of tenement or terraced housing • Lack of greenery and open space	• Department shops • Large offices • Main railway and bus stations • Many churches, pubs, clubs and cinemas and the town hall • Easily accessible - the main roads head into the C.B.D. • Highest land value • High building density • Tall buildings • Traffic congestion • The oldest part of the city • Grid iron or medieval street pattern • Subject to constant change	• High rise flats • Derelict land and waste ground • Derelict land, vandalism, crime, poverty, unemployment and other social and economic problems – related to industrial decline • Redevelopment occurring • Environmental improvements exist in some areas • Mixture of old and newer buildings • New houses of mixed type • Recent regeneration to encourage more people to live there	• Well-planned • Street patterns of crescents and cul-de-sacs • Some outer-city council estates with flats, high rise and mixed housing • Growth of car ownership, commuting and demand for quieter and safer residential environments have led to the growth of these areas • Houses with garages	• Room for expansion • Industrial estates • New housing estates • New business district e.g. retail parks and shopping centres

Fig 2.12 A city in the concentric land use model

Look at how buildings change in different zones of land use models.

Fig 2.13 Sydney CBD

Fig 2.14 Bangkok inner city

Fig 2.15 Suburbs of Arizona

Quick Test

1. What is a conurbation?
2. Describe the different kind of buildings you might find in the suburbs, the inner city and the C.B.D.

Recent developments or changes in a developing world city

The rapid growth of cities in developing countries – rural to urban migration

You must also learn a specific case study related to a developing world city and be able to apply it to questions in the SQA exam. Cities in the developing world have grown rapidly in the past 25–30 years.

The movement of people from the countryside (rural) to towns and cities is called rural-urban migration. This movement is due to rural push factors and urban pull factors. Often the pull factors are greater than the push factors. Rural families perceive that the city offers greater opportunities than the countryside, but the reality is often very different. Migrants arrive without money and are often homeless, living in temporary shelters made of waste materials.

Urban pull
Better job opportunities
Higher potential earnings
Better health and education services
Better amenities and infrastructure

Rural push
Poor job opportunities, low pay
Poor health and education services
Less access to service and industrial jobs

Fig 2.16 Urban pull and rural push

Urban land use within cities in developing countries

As in cities in the developed world, cities in developing countries acquire recognisable patterns of land use over time. These differ from the patterns in developed countries in a number of ways:

- High quality housing is located nearest to the C.B.D. (1)
- The quality of housing decreases rapidly towards the edge of town where slums or shanty towns develop (4)
- Industry is located along main roads leading into the city centre (6)

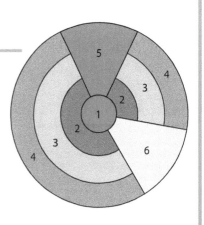

Case study – Rio de Janeiro

Rio de Janeiro is the second largest city in Brazil. It is an important centre of business and industry but is also a city of contrasts with old and new buildings, motorways and dirt tracks, and shacks and skyscrapers sitting side-by-side.

Rio is divided into five zones – Centro, North Zone, South Zone, West Zone and Barra da Tijuca. The photograph here shows the statue of Christ. His left arm is pointing towards the North Zone and his right arm to the South Zone.

Fig 2.17 Christ the Redeemer, Rio de Janerio, Brazil

Problems in developing cities – Rio

In most cities within developing countries there are extreme contrasts between housing for the rich and housing for the poor who are forced to live in shanty towns (favelas) on the edge of the city, such as Rocinha, shown here.

Fig 2.18 Rocinha: a shanty town (favela) in Rio

Problems in Rio

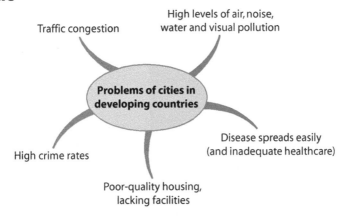

Traffic congestion

High levels of air, noise, water and visual pollution

Problems of cities in developing countries

High crime rates

Disease spreads easily (and inadequate healthcare)

Poor-quality housing, lacking facilities

Solutions to problems in Rio

One-third of the people living in Rio live in squatter settlements. Local authorities cannot afford to remove and replace them so they try to improve existing housing and have introduced self-help schemes:

1. Self-help schemes in Rocinha.

2. New transport routes, e.g. tunnels through the solid crystalline Serra do Mar mountain.

3. Building of new suburbs, e.g. Barra de Tuca, to create more housing.

4. Improved primary health care.

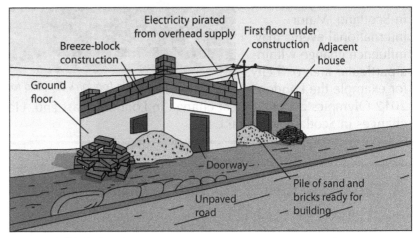

Electricity pirated from overhead supply

First floor under construction

Adjacent house

Breeze-block construction

Ground floor

Doorway

Unpaved road

Pile of sand and bricks ready for building

Fig 2.19 Self-help housing

5. Long term, the best solution is to reduce migration to the cities by helping to spread development and wealth to rural areas; although this has proven almost impossible to achieve.

6. Local authority site and service schemes.

7. Charity organisations provide money for improved housing and education programmes.

Quick Test

For any named developing city you have studied, describe measures taken to improve the quality of life in shanty towns.

Recent developments or changes in a developed world city

You must learn a **specific case study** in detail and be able to apply it to questions in the SQA exam. Look at each sector of your named city and ensure you have named examples for any recent (within 30 years) changes.

You may choose any world city, not just in Scotland. Major international events may influence change within a particular area of a city, for example the London

Fig 2.20 Construction in London's East End for the Olympics

2012 Olympics actioned great change in London's East End. However, here we look at changes in Scotland's central belt.

A tale of two cities: urban change in central Scotland

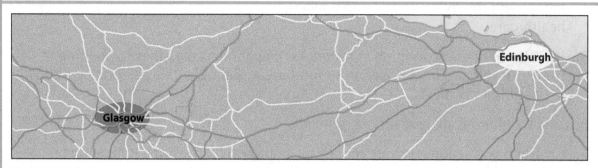

Fig 2.21 A tale of two cities: urban change in central Scotland

Glasgow — change in the C.B.D.	Recent developments in the rural/urban fringe	Edinburgh — change in the inner city

• Pedestrianisation, e.g. in Buchanan Street and Argyle Street
• Upgrading of C.B.D. open space, e.g. George Square
• Diversification of city employment – a much greater emphasis on the tourist industry (significance of city break holidays) leading to increased bed accommodation in new C.B.D. hotels (Hilton, Radisson)
• Alteration of C.B.D. road network – one-way streets (around George Square), use of bus lanes to discourage use of private transport and encourage use of public transport
• Renovation and redevelopment of many C.B.D. sites to provide modern hi-tech office space (e.g. Lloyds TSB, Direct Line etc.)
• 24/7 use of the city centre (e.g. Cineworld in Renfrew Street)
• High-profile shops to attract more customers, e.g. the Apple Store
• New covered shopping centres, e.g. Buchanan Galleries
	• As both Edinburgh and Glasgow experience urban sprawl, the cities are growing across the central belt
	• Recent changes include:
	  – New housing estates
	  – Out-of-town shopping complexes e.g. Straiton or Braehead
	  – Park and ride complexes
	  – Development of tertiary and quaternary industrial estates, e.g. RBS at Gogarburn

		• Decline of old industry → areas cleared for new tertiary industries, e.g. leisure complex Fountain Park
		• Redevelopment of brownfield sites, e.g. Sainbury's between Gorgie Road and Westfield Road
		• Environmental landscaping – e.g. Canning Street
		• Clean up of Edinburgh Canal – now used for leisure e.g. Re-Union Canal Boats
		• Gentrification of area around the Union Canal – new luxury flats, new cafés and restaurants opened, e.g. Zizzi at Edinburgh Quay
		• Regeneration of housing, e.g. Gardner's Crescent
		• Improved transport links e.g. the Western Approach Road

EXAM TIP

Look at each sector of your named city and ensure that you have named examples for any changes that have happened within the last 30 years.

Quick Test

1. List five recent changes that have taken place in an inner city area of a city you have studied.

2. For a named city you have studied in a developed country, give reasons for the main changes that have taken place in the city centre shopping area over the last 30 years.

Changing countryside in the European Union (EU)

There are over 500 million people living in the EU. 75 million live and work in rural areas, making farming and rural industries a key part of the economy. The EU produces huge quantities of agricultural and other primary products. Many millions of people depend on the land, or on processing land-based products, for a living. Foodstuffs account for 6.8% of EU exports at a value of over 116 billion euros.

EXAM TIP

Make use of detailed diagrams like the one in this section by extracting the information contained in the annotations. Converting the information into another form, such as a flow chart or before and after list, is a good method for learning and revising for exams.

Le Midi in the past

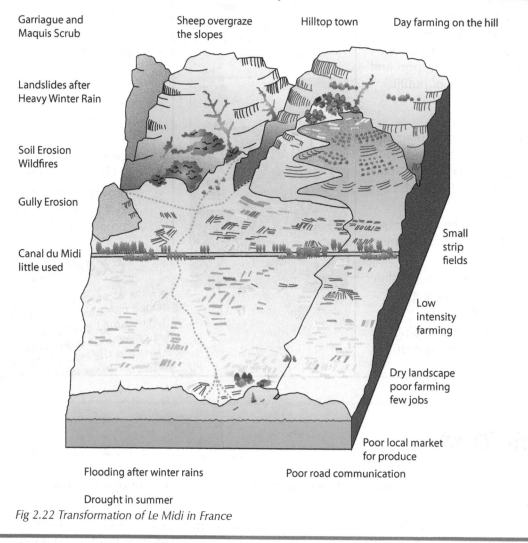

Garriague and Maquis Scrub

Sheep overgraze the slopes

Hilltop town

Day farming on the hill

Landslides after Heavy Winter Rain

Soil Erosion Wildfires

Gully Erosion

Canal du Midi little used

Small strip fields

Low intensity farming

Dry landscape poor farming few jobs

Poor local market for produce

Flooding after winter rains

Poor road communication

Drought in summer

Fig 2.22 Transformation of Le Midi in France

How is farming changing in the EU

The diagram shows changes in the landscape of Southern France and exemplifies how many rural areas in developed countries (like those in the EU) have been through a great transformation from old-style farming to modern agribusiness. People living in rural areas have seen their economy improve along with roads and towns. They have benefited from better-performing farming industries that are able to move goods to distant markets and from new industries that provide local jobs and improve the economy.

DID YOU KNOW?

Fewer and fewer people live in the countryside in Europe. This is called rural depopulation. Rural areas experience poorer economic conditions than towns and have less social and transport infrastructure.

Le Midi today

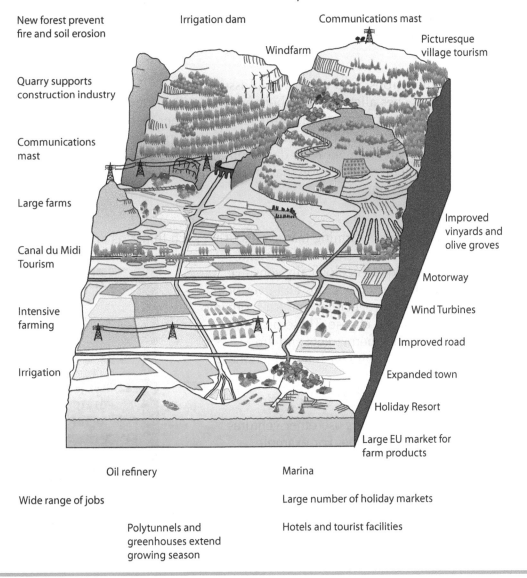

New forest prevent fire and soil erosion

Irrigation dam

Communications mast

Windfarm

Picturesque village tourism

Quarry supports construction industry

Communications mast

Large farms

Canal du Midi Tourism

Intensive farming

Irrigation

Improved vinyards and olive groves

Motorway

Wind Turbines

Improved road

Expanded town

Holiday Resort

Large EU market for farm products

Oil refinery

Marina

Wide range of jobs

Large number of holiday markets

Polytunnels and greenhouses extend growing season

Hotels and tourist facilities

Modern factors affecting farming

There are a number of factors at work that affect modern farming.

The Common Agricultural Policy (CAP)

This is a policy across the EU that supports farmers and helps to maintain stable prices to guarantee the income that farmers get for their products. EU policy also regulates aspects of farming, such as animal welfare, the use of genetically modified (GM) crops and what chemicals and fertilisers are safe to use on farms. For example, the EU has banned certain practices, such as keeping young cattle in veal crates. Farmers use set-aside land to prevent over-production of certain crops in return for guaranteed prices and payments.

GM crops

Genetically modified crops are a controversial issue. Manipulating plant genes allows certain beneficial characteristics of one organism or plant to be transferred artificially to another (genetic modification). This can increase crop yields or improve resistance to disease, pest or drought. Many people object to this arguing that it might have dangerous effects that impact on the environment and the effects of genetic modification may 'escape' and spread into non-modified crops or the the natural environment.

Organic farming

This is a specialised form of farming that avoids using such things as artificial fertilisers and pesticides and other agri-chemicals. Instead, more traditional methods or ecologically sensitive techniques are used to produce crops and livestock products such as organic vegetables and fruit, organic beef and organic milk and cheese. These products attract a higher price in supermarkets and are favoured by many people who wish to reduce their impact on the environment.

Fig 2.23 Farm shop

Diversification

Diversification is when farmers use other ventures to support their farming income. For example, some farms set up farm shops and restaurants and/ or have visitor areas. Other farms may have tourist-related activities or outdoor activities, such as off-road driving. Farms may lease holiday cottages or provide 'on the farm' accommodation. Wind farm development may also generate extra income.

Biofuels

One way of generating new income is to produce biofuels, which are used in special power stations or in regular generating stations as a substitute for fossil fuels. Farmers may grow fast-growing trees, such as willow, which can be cropped, or may grow oil-producing plants for bio-diesel. This industry allows farmers to diversify and also to be involved in carbon capture schemes favoured by governments anxious to reduce the effects of global warming.

New technology

The diagram of Southern France on page 67 shows how farming has become increasingly mechanised and uses much more control in the environment. Computerised systems, such as the use of GPS for managing field operations or for managing animal feeding, are typical of modern hi-tech farming. Production can be increased using water management systems that provide irrigation and drainage and through the use of poly-tunnels and environmental control systems.

Fig 2.24 New technology helps improve efficiency on farms

Government policy

Governments regulate and support their farming industries. In the UK, for example, the Department for Food and Rural Affairs (DeFRA) supports farming and rural communities. The Scottish Rural Development Programme does a similar job. Governments regulate farming practices and provide subsidies and support for farmers and rural industries. For example, in the UK there are laws covering animal welfare and the use of methods like GM crops. Fuel for farm vehicles is taxed at a lower rate. Governments provide money for research into agriculture. Disease control is mandatory, to prevent diseases such as foot and mouth disease, bovine TB and plant pests and diseases such as ash dieback and Colorado beetle infestation of potatoes.

> **EXAM TIP**
>
> You should know and understand the changes in the rural landscape in developing countries (e.g. India) related to modern developments in farming such as:
> • genetic modification (GM)
> • impact of new technology
> • biofuels

Quick Test

1. What is the CAP of the European Union?
2. What are the disadvantages of using GM crops and why do some governments ban the import of GM crops from other countries such as the USA?
3. How can modern farming protect the environment?
4. How is foot and mouth disease or bovine TB controlled in Britain?

Brazil developing the Amazon

Brazil has a population of over 200 million people, with about 25 million living in the Amazon Region. In 1960, only 2 million people lived in this challenging environment so there is clearly major population pressure on the Earth's largest area of surviving equatorial rainforest.

What are the pressures on Amazonia?

Brazil currently has a National Income per head of $11,000 which is low compared to developed countries like the UK, where it is almost $36,000. Wealth in Brazil is not equally distributed across the country. The rich south-east region, dominated by large industrial cities and the rich agricultural heartland of the nation, generates over 56% of the national wealth. The poorest region is the north-east (sometimes referred to the Sertao), and contains the poorest district of all – the Caatinga. Brazil has a national policy to try to develop the country's resources, particularly those in the underdeveloped Amazonian region.

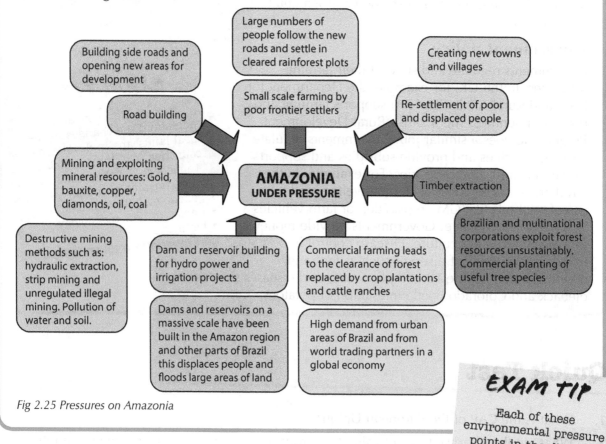

Fig 2.25 Pressures on Amazonia

EXAM TIP

Each of these environmental pressure points in the diagram would be a good topic for an investigation of the problems facing the Amazon rainforest.

The pressures diagram shows some of the main environmental pressure points facing Amazonia. With 20% of the rainforest cut down and being exploited, the threat to the environment has drawn international attention. These figures, calculated from satellite observations, show the pace of change:

Forest degradation 2007–2010 (Thousands of Hectares)				
	2007	**2008**	**2009**	**2010**
Amazonas	257	412	181	459
Mato Grosso	8951	12 987	8486	2502
Para	3899	8264	1558	3499
Rondonia	412	643	232	315
Amazonia	15 983	27 413	13 301	7508

Why is the Amazon under pressure?

The pressures have been created by the need for development to improve the economy and to help poor landless or displaced people from areas like the drought-affected north-east Brazil, as well as population growth. New settlers move into the region using the new roads. With or without government approval, they exploit forest timber reserves and make a living. Some people are given government land grants, while others who have migrated in as squatters work as subsistence farmers on small plots of cleared forest land to grow crops and keep animals.

Large dam-building projects like the Tucurui Dam or the new Belo Monte Dam displaces people, and those in other areas like the Sertao of the north-east region, have been displaced and encouraged to migrate to Amazonia.

Large mining companies are given licences by the government to prospect and exploit mineral and energy resources in the region. This supplies the Brazilian economy and generates export income. Similarly, multinational companies wish to exploit timber resources and ship them to developed countries to provide building materials and paper supplies. Multinational and national food processing companies use cleared land to grow cash crops such as coffee, fruit or palm oils. Cattle ranching helps meet the demand for meat and processed meat products like corned beef sold in supermarkets.

The growing south-east region and its towns and cities need cheap electricity for industry and for domestic supply. The Amazon River and its major tributaries provide this, but at the cost of losing forest areas to flooding.

Quick Test

1. Which region of Brazil shows the greatest amount of rainforest clearance?
2. Which is the poorest region of Brazil?

Sustainable development in Brazil

Can the rainforests be developed sustainably?

In the last 100 years, whole areas of the world have seen their forests cut down and lost to farming, urban development and industrial use. Population pressure in Indonesia, Malaysia and large areas of South-East Asia have seen their forests disappear, placing animals like the orangutan and other wildlife onto the endangered species list. Forest is replaced by industrial plantations producing food crops for export, industrial crops like rubber, or oil palm for food processing and industry. Urban development in industrialising countries spreads onto forested areas.

African forests have been similarly exploited and the rich forest environment may be replaced by scrub and secondary growth. This has had a major impact on wildlife, for example on the island of Madagascar where habitat destruction has resulted in many of its unique wildlife species, such as lemurs, being threatened with extinction. In India and Pakistan the remaining tiger population is now largely confined to natural reserves that are also under pressure from farmers and poachers. Without policies for preserving and developing forest resources sustainably, there is a chance that even the remaining forest areas have a limited future.

Why should equatorial and tropical forests be preserved?

Maintaining a healthy planet requires biodiversity. Natural systems tend to work on a feedback loop system so that upsetting the balance of nature is likely to have consequences down the line. Large-scale forest clearance by burning causes increased carbon dioxide levels in the atmosphere and contributes to climate change. Managing forests by clearing and thinning them and allowing secondary growth can cause large-scale bush fires, which in turns creates pollution of the atmosphere. Fires also threaten newly developed urban and industrial areas.

Fig 2.26 Forest clearance in the Amazon rainforest

Globalisation means that demands for supermarket foods and confectionary generates conflict as conservationists and local people object to new areas being developed for commercial farming. Conflicts are played out through the media and on the Internet as even small campaigns can undermine products. For example a campaign aimed at Nestle's Kit Kat because of oil palm planting in Indonesia was effective in changing company and government policy.

Major projects such as dams, mining schemes and agricultural developments do not always favour local people. Many are displaced and lose their land, jobs are often fewer than anticipated and go to incomers and people with technical skills. Other jobs are often unskilled and low paid and lead to exploitation of workers. Alternative opportunities, such as eco-tourism development or local sustainable schemes based on forest resources, are lost to multinational corporations or government agencies with dubious green credentials.

Forest area resources in remote districts can lead to political instability as rebel groups target mining sites to exploit mineral resources and finance revolutionary activity. Sierra Leone, Liberia, Rwanda and the Congo have all experienced this type of exploitation, and blood diamonds or illegal gold find their way onto the international markets. This often finances further terrorism and denies countries development money.

Sustainability options

The solution to the development versus rainforest conservation argument is difficult when it is clear how important development is to the Brazilian people, particularly the poor. The Brazilian Government has created a national parks authority, IBAMA, which not only administers 50 national parks but also looks after a wide range of natural, heritage and ecological reserves.

Carbon capture/replacement schemes have been put in place to develop long-term use of the forests. Many world conservation bodies, partner governments and commercial organisations have invested in rainforest conservation in Brazil. WWF and other agencies such as the UN environment programme have been involved in projects aimed at developing conservation strategies.

Major developments have met with international protests. The Monte Belo Dam project has been challenged in the Brazilian courts; conservationists have questioned the role of the World Bank in financing schemes like this that are likely to cause destruction of the natural environment. Indigenous peoples and local residents form pressure groups that raise concerns in Brazil and abroad. Consumer pressure can be effective in encouraging sustainable production. Fair trade schemes, sustainable source schemes for tropical timber and international agreements such as the CITES restrictions on the animal and animal products trade all contribute to a more sustainable future.

Quick Test

1. What is a plantation?
2. What are blood diamonds?
3. Why is it important to restrict the trade in rare animal products?
4. Why is fair trade a good thing?

Investigating Brazil's environmental issues

Investigations

At many points in your National 5 Geography course you will be expected to conduct investigations or research essays and presentations. You will be given options for topics to study and follow at more depth. The rainforest and conservation issue in Brazil is one of international importance because these forests are the largest expanse of this natural system remaining in the world, which makes this fertile territory for investigations.

Making a start – sample where, why, what, when, how questions

1. Draw up a list of key questions about Brazil and its rainforests to develop your background knowledge of Brazil.

2. Collect and study maps, diagrams and images of Brazil: its cities and its natural environment.

3. Collect development statistics and others to compare against UK statistics.

4. Find 10 'did you know facts' about Brazil and Amazonia.

5. Design a series of search questions and research phrases relating to conservation and environmental issues. Find what issues are going on today. Look at the effects of past projects to find out whether they were successful.

6. Ask questions that relate to people affected by developments.

7. Identify conservation bodies working in the field.

Fig 2.27 Rio de Janeiro, Brazil

8. Investigate alternatives to proposed developments and follow up conservation schemes run by charities and other agencies.

Laying out a presentation

EXAM TIP

- Use a wide range of sources in investigative work rather than relying on one or two.
- Try to use complex sources as well as simple ones.
- Don't always go for sources at the top of the list.
- Cut and paste useful sections into a collecting document – be sure to note the source for later!
- Acknowledge your sources, and indicate if you have adapted a source as you use it, to avoid plagiarism.

Presentations take many forms in Geography. One thing they have in common is the use of maps, diagrams, photos and graphic items. Collecting a range of these will enable you to add visual interest to your presentation but will also provide valuable resources to support your analysis of the issue. The following list gives a few suggestions.

- Maps: political, physical, climate, vegetation, economic activity, road and transport network, population density.
- Diagrams: model diagrams, cross sections, rainforest structure.
- Photographs: projects, satellite images, people involved, companies.
- Graphs: climate; population and development statistics; regional data for cross comparison; showing change over time.

Once you have a range of items work out their logical order and how you intend to use them in the presentation or report. All graphics should be processed and added to for added value. This means re-drawing and adapting, labelling and annotating, and turning raw data into meaningful graphs.

Presentation structure

Plan a logical structure for your presentation. Work out what you are trying to tell your audience and a logical sequence. If you are trying to be convincing, build to a conclusion using logical steps to be sure your audience stays with you. Consider how and when you use graphics to support your case. Develop initial impact by designing a short snappy introduction. Lay out you argument in advance. Present a balanced view by examining and criticising the arguments of both sides. Come to conclusions along the way and give your reasons for them. Summarise at the end the main points and draw your main conclusion.

Write up you first draft and read it over so you can edit it. Ask the key question (is this my own work?). Complete your final draft and check it over for remaining typos and mistakes.

What is climate change?

Climate change refers to a significant and lasting change in long-term weather patterns over periods of time ranging from decades to millions of years. It may be a change in average weather conditions, or in the distribution of weather statistics around the average conditions (i.e. more or fewer extreme weather events). Climate change is often confused with global warming as there has been a general increase in average global temperatures in the last 150 years.

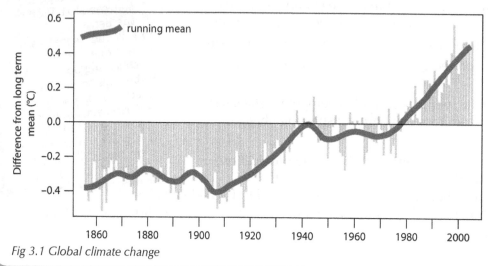

Fig 3.1 Global climate change

Evidence for climate change

EXAM TIP

You can find useful information on climate change on the Met Office website: http://www.metoffice.gov.uk/climate-change/guide/what-is-it.

- Long-term evidence (1000s years): ice core analysis.
- Medium-term evidence (100s years): historical records, glacial retreat, dendro-chronology.
- Recent evidence (decades): changes in biodiversity, climate records, sea ice limits.

Fig 3.2 Melting glaciers, Iceland

Causes of climate change

Greenhouse gases are produced naturally but are added to by human activity. This causes the **enhanced greenhouse effect,** which is believed to be the main reason behind the current pattern of climate change.

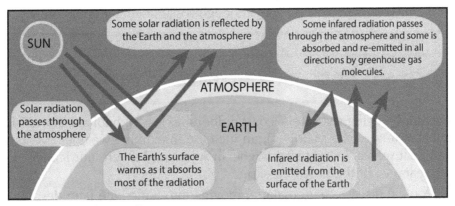

Fig 3.3 The greenhouse effect

Physical (natural) causes	Human (anthropogenic) causes
Changing output of solar radiation – changes in the amount of energy given out by the sun over time.	**Fossil fuels** – Increased consumption of coal, oil and natural gas has increased the amount of CO_2 in the atmosphere.
Volcanic eruptions – and the amount of volcanic dust particles which enter the atmosphere as a result. Dust acts as a blanket shielding the Earth from incoming solar radiation and leads to a decrease in temperature.	**Deforestation** – especially the burning of tropical rainforests adds to the concentration of CO_2 and decreases the amount used by plants in photosynthesis.
Wobble roll and stretch theory (the Milankovitch Theory) – changes in the movement of the Earth's angle of tilt, its orbit around the Sun and the gravitational pull of the Sun and the Moon all change the amount of energy the Earth receives from the Sun, therefore influencing climatic temperature.	**Increased methane** – this comes from flatulent cattle, rice padi fields and waste in landfill sites, which are all caused by an increase in the population.
	Increased output of nitrous oxides – these are from agricultural fertilisers and pesticides.
Changing oceanic circulation – for example, **El Nino**/La Nina.	**CFCs (chloroflourocarbons)** – these leak from old fridges and can remain in the atmosphere for up to 400 years. They add to the enhanced greenhouse effect.

EXAM TIP

Refer to specific greenhouse gases and how they are produced.

Quick Test

1. Describe the physical factors which have led to climate change.
2. Explain how human activity has influenced climate change.

Consequences of climate change

Many effects of climate change are negative and will have the greatest effect on developing countries and the world's poorest people, although some potential impacts are positive.

Negative effects of climate change

- **Environmental** – rising sea levels, coastal flooding, loss of habitats, some islands uninhabitable (e.g. in the Maldives), an increase in extreme weather events such as European heat waves, increased droughts in Africa and severe hurricanes in the USA, the disappearance of the polar ice cap.
- **Social** – tropical diseases (e.g. malaria) moving into temperate latitudes, climate wars fought over precious food and water.
- **Economic** – a loss of locally important economic activities such as skiing in the Alps.

Fig 3.4 Hurricane Sandy caused more than $68 billion worth of damage in October 2012. Increased hurricanes can be an effect of climate change

Positive effects of climate change

Social – Warming of northern latitudes leading to increased tourism, for example, warmer Scottish summers.

Economic – Changing agricultural patterns e.g. maize grown in Scotland instead of potatoes; improved crop yields as warmer temperatures ensure longer growing periods. This can lead to a reduced level of dependency on other nations if countries become self-sufficient in terms of food supplies.

> **EXAM TIP**
>
> Make reference to positive and negative effects on the economy and the environment in both developed and developing countries. Use up-to-date named examples.

Fig 3.5 Warmer temperatures improve crop yields

Reducing the effects of climate change

You should be able to distinguish between strategies that lead to both prevention (avoiding climate change) and mitigation (making the effects less severe).

- **Local strategies**, such as 'Reduce, Reuse, Recycle' (household recycling), reducing food miles, cycle to school and switching off lights when not in use.
- **National strategies** (government initiatives), such as road tax for 'gas guzzlers', low-emission vehicle incentives and carbon reduction initiatives in homes such as boiler upgrades, loft insulation and solar panels.
- **International strategies**, such as geo-engineering, improved technology, United Nations agreements/regulations concerning climate change, the Intergovernmental Panel on Climate Change (IPCC).

Quick Test

1. How can we reduce the effects of climate change?
2. What evidence is there that climate change is taking place?

Equatorial and tundra environments

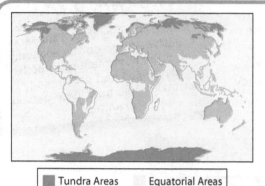

Tundra Areas ▮ Equatorial Areas ▯

Fig 3.6 Equatorial and Tundra regions

Equatorial (rainforest) environments

- **Location:** these are dense rainforest areas found in lowland areas 10 degrees either side of the Equator (0)

- Often described as the 'world's lungs' because of all the oxygen produced by the forests

- **Regions/countries with an equatorial rainforest climate:** Amazonia in Brazil, Peru, Indonesia, Colombia, Democratic Republic of Congo

- **Climate summary:**

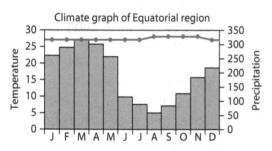

Climate graph of Equatorial region

▯ Precipitation (mm) ●– Daily Mean (°C)

- Total annual precipitation: 2286 mm

- Maximum temperature: 27°C

- Minimum temperature: 26°C

- Range of temperature (remember the difference between the max. and the min. temp): 1°C

- It is hot and wet all year round

Tundra environments

- **Location:** these are cold desert areas mostly found north of the Arctic Circle (66½ N)

- In Finnish, tundra literally means 'barren or treeless land'

- **Regions/countries with a tundra climate:** Northern Alaska, Northern Canada, Northern Russia (Siberia) and Northern Scandinavia

- **Climate summary:**

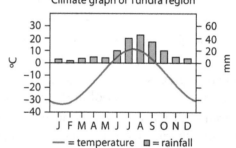

Climate graph of Tundra region

—— = temperature ▯ = rainfall

- Total annual precipitation: 250 mm

- Maximum temperature: 14°C

- Minimum temperature: –28°C

- Range of temperature: 42°C

- Very cold, dark, long winters

- Very short, light summers

- The amount of rain and snow are so low (<250 mm a year) that the tundra is classed as a desert

Equatorial vegetation

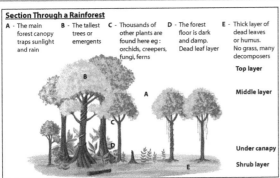

Section Through a Rainforest

A - The main forest canopy traps sunlight and rain
B - The tallest trees or emergents
C - Thousands of other plants are found here eg : orchids, creepers, fungi, ferns
D - The forest floor is dark and damp. Dead leaf layer
E - Thick layer of dead leaves or humus. No grass, many decomposers

Top layer

Middle layer

Under canopy

Shrub layer

Fig 3.7 Equatorial vegetation

- All the trees are broadleaved evergreens because the high temperatures and evenly distributed precipitation through the year allow continuous growth.
- The canopy is a dense layer of leaves allowing very little light to penetrate to the forest floor. This canopy has an abundance of flowers and fruit.
- Buttress roots support the very tall trees or 'emergents' (50–60 m high).
- Without the rainforest canopy the soils become infertile very quickly as the heavy rainfall soon leaches all the nutrients.

Tundra vegetation

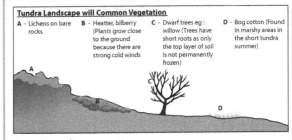

Tundra Landscape will Common Vegetation

A - Lichens on bare rocks
B - Heatter, bilberry (Plants grow close to the ground because there are strong cold winds)
C - Dwarf trees eg : willow (Trees have short roots as only the top layer of soil is not permanently frozen)
D - Bog cotton (Found in marshy areas in the short tundra summer)

Fig 3.8 Tundra vegetation

- The growing season is very short (about 2 months).
- In the short summer only the surface soil thaws (top 50 cms) and the subsoil remains frozen (permafrost).
- This leaves the soil waterlogged in summer and there it has a lack of nutrients.
- The vegetation has to survive extreme cold, biting winds and a moisture deficit.

Equatorial wildlife

- Animal and bird life in the rainforest are abundant due to the availability of food and shelter.
- Equatorial rainforests are the most diverse ecosystem on Earth and have two thirds of all the world's wildlife species, e.g. monkeys, sloths, jaguars, parrots.

Tundra wildlife

- The tundra has fewer species of plants than any other ecosystem.
- Conditions are very difficult for the species that live here, e.g. caribou, arctic wolf, musk ox.

Quick Test

1. Which two climatic characteristics mean that there is continuous growth in equatorial rainforest environments?
2. Why are tundra areas classed as deserts?
3. What is the direct result of removing the rainforest canopy?
4. Where are equatorial and tundra environments located?
5. How does the climate influence vegetation in equatorial and tundra environments?

Misuse of equatorial rainforests

The broken nutrient cycle

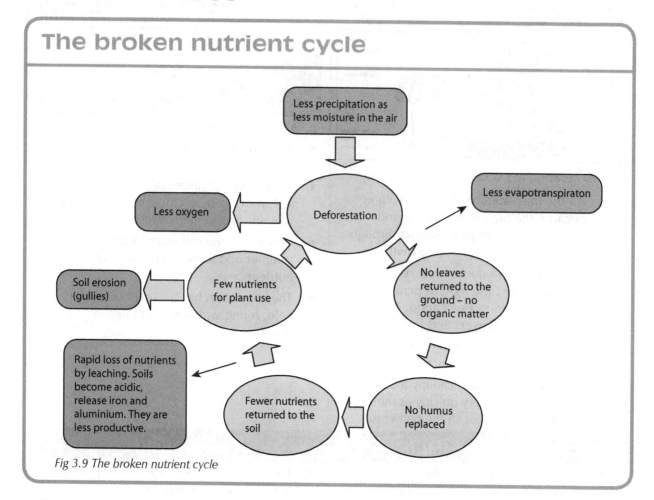

Fig 3.9 The broken nutrient cycle

Causes of deforestation

- **Logging** – large areas of rainforest are being clear felled for hardwoods, such as mahogany and rosewood.
- **Plantation farming** – areas are cleared for the cultivation of cacao, soya and coffee.
- **Cattle ranching** – large areas are cleared for the production of beef. Most ranches only last 10 years as the grass quality decreases because of the poor soils.
- **Construction of hydroelectric power (HEP) schemes** – large areas of the forest are flooded for the generation of electricity.
- **Roads/communications** – the population of Amazonia is growing and so are the number of developments and these all need access by a network of roads.
- **Mineral extraction** – Amazonia is rich in mineral deposits, e.g. gold, bauxite and iron ore, which attract mining companies.

Effects of deforestation

- With a lack of tree cover, run-off and soil erosion increases.
- This leads to gullying, flooding, landslides and increased sedimentation in rivers.
- Habitat loss means the reduction of the number of animal and plant species (the biodiversity) able to survive in the rainforests.
- Loss of an invaluable resource of raw materials for medical drugs.
- If the rainforest is cleared by burning, the carbon dioxide released contributes to global warming.
- The trees convert carbon dioxide into food and release oxygen; with fewer trees this process is less effective.
- The indigenous Indians lose their lands and way of life as they are forced to live in a smaller area of rainforest.

Possible solutions to deforestation

- **Banning logging** – this is difficult to enforce as the forest is very dense and difficult to patrol. It can just lead to the smuggling of illegal timber.
- **Different logging methods** – this is more realistic. Selective logging means only trees with a certain thickness and the least important trees are felled. There are also controls on the use of heavy machinery.
- **Satellite imagery** – vast areas can be monitored with satellite imagery to show the extent of deforestation. However, not all countries have the money or technical knowledge to implement this technology.
- **Conservation and replanting** – this can be difficult on the poor-quality soils, and competition from grasses mean the tree seedlings often die.
- **Certification** – this involves marking the wood with a barcode so that the wood can be sourced from 'stump to store shelf'. This means customers can make informed choices and choose to buy sustainable, or 'eco-friendly', wood.

Quick Test

1. What is deforestation?
2. What impact does rainforest deforestation have on medical advances worldwide?
3. Why might trying to impose a worldwide ban on logging in equatorial rainforests be difficult to enforce?
4. Identify four causes of deforestation.

Misuse of the Tundra

Although the Tundra is a region of low population density it has many natural resources. Many developed countries would like to exploit these resources and in Alaska the Americans are already busy doing so, but as you can see from the articles below this is often at the expense of the natural environment:

Alaska hit by 'massive' oil spill

On March 2nd 2006 an oil spill was discovered on the Prudhoe Bay oil field. Experts estimated that up to 267 000 gallons (one million litres) of crude oil had leaked from a corroded pipeline in the northern part of the Alaskan state. The spill was described as 'a catastrophe' by environmentalists. Alaska was still recovering from the Exxon Valdez shipping disaster in 1989 when 11 million gallons

(42 million litres) of oil was washed on to the Alaskan coast.

PROPOSED PEBBLE MINE, BRISTOL BAY, ALASKA

In October 2012 a proposal by a transnational corporation for a copper and gold mine was put forward. This region is an area of fragile tundra on the shores of Bristol Bay. It will also be close to two of Alaska's beluga whale populations including the last 340 belugas of the Cook Inlet. Environmentalists have demanded that the mining giant, Anglo American Corporation, leave Bristol Bay and the beluga whales alone. Then an Environmental Impact Assessment (completed in April 2013) has also claimed that the toxic waste that would be produced by the mine and released into the Bay would directly threaten the salmon farming industry which is so important to the Alaskan economy.

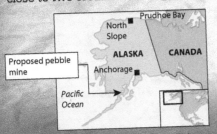

Fig 3.10 Effects of the proposed pebble mine in Alaska

Effects on the tundra landscape

- Disturbance of the delicate ecosystem
- Disruption to wildlife habitat and migration patterns, e.g. caribou migration
- Pollution of the land and water courses
- Abandoned waste
- Destruction of one of the last of the world's wilderness areas

The tundra is a very fragile ecosystem in delicate balance. Once disturbed by human activity it may take many years to re-establish.

Positive effects of the proposed pebble mine	Negative effects of the proposed pebble mine
• Creation of jobs. • Provide tax revenues for the state of Alaska. • Reduce dependence of the US on foreign sources of raw materials.	• Decline of fish populations. • Contamination of sea water by waste water. • Large amounts of waste rock material. • Giant open cast mine cut into the tundra landscape. • Disruption of local drainage basins and river flow. • Disruption to wildlife in this currently wild and undeveloped landscape. • Beluga whale population under threat. • The development will require the building of miles of roads and bridges.

Quick Test

1. Name two activities that threaten tundra environments.
2. Identify three positive effects of development on the tundra environment.

Environmental hazards

In this section you must know:
- the main features of earthquakes, volcanoes and tropical storms
- the causes of each hazard
- the impact of each hazard on people and the landscape
- how to manage each hazard: methods of prediction and planning, and strategies adopted in response to environmental hazards.

For each of the hazards, you should have a case study example to refer to in examination answers.

What are environmental hazards?

Natural hazards come in many forms and are different from natural disasters.
- Natural hazards are either geomorphological occurrences relating to floods, earthquakes, volcanoes, landslides, tsunami or meteorological phenomena such as hurricanes, storms, floods, forest fires and droughts.
- Natural disasters occur when people are overtaken by the natural hazards listed above. This will involve loss of life, injury and destruction of property. It will have an impact on future economic development and prosperity that extends beyond the immediate effects of the disaster involved.

The Earth's structure

The Earth consists of four layers: **inner core, outer core, mantle** and **crust**.

The mantle can be subdivided into the lower and upper mantle.

The upper mantle along with the crust makes up the lithosphere.

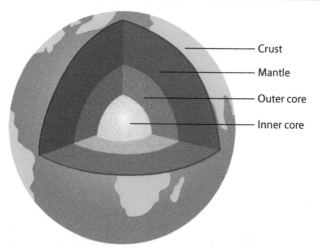

Crust
Mantle
Outer core
Inner core

Fig 3.11 The structure of the Earth

Tectonic plates

The Earth's crust is made up of a number of plates – solid rock which float on the mantle.

Some plates are formed of continental crust – these are the land masses, whereas others are formed of oceanic crust – these are the ocean floors.

Plates move relative to each other. At the plate boundaries, plates may move away from, towards or sideways past neighbouring plates. There are then *three* types of plate margins or boundaries.

Type of Plate Boundary	Key Information
1. **CONSTRUCTIVE**	• Direction of plate movement: Apart • What happens to the crust? Crust separate and new crust forms • Named Example: Mid Atlantic Ridge • Eurasian and North American plates • Features found: Earthquakes, Volcanoes, New land
2a. **DESTRUCTIVE SUBDUCTIVE** **(subduction zone due to oceanic and continental crust)**	• Direction of plate movement: Towards • What happens to the crust? Destroyed • Named Example: Nazca and South American plates • Features found: Earthquakes, volcanoes, mountains, ocean trench
2b. **DESTRUCTIVE COLLISION** **2 x continental crust**	• Direction of plate movement: Towards • What happens to the crust? Crumples • Named Example: Indo-Australian and Eurasian Plates created the **HIMALAYAS** • Features found: Earthquakes, mountains
3. **CONSERVATIVE** or	• Direction of plate movement: Alongside • What happens to the crust? It is conserved, ie, no land is created and no land is destroyed • Named Example: San Andreas Fault • Features found: Earthquakes

Types of plate boundaries

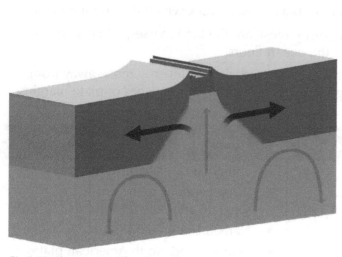

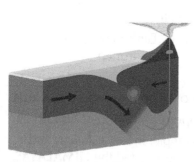

Fig 3.12 Constructive Plate Boundary

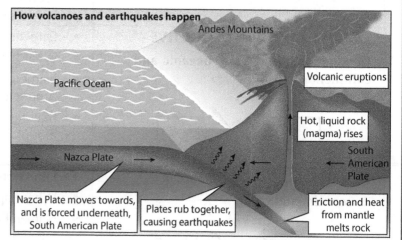

How volcanoes and earthquakes happen

Andes Mountains

Pacific Ocean

Volcanic eruptions

Hot, liquid rock (magma) rises

South American Plate

Nazca Plate

Nazca Plate moves towards, and is forced underneath, South American Plate

Plates rub together, causing earthquakes

Friction and heat from mantle melts rock

Fig 3.13 Destructive Plate Boundary

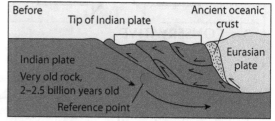

Before
Tip of Indian plate
Ancient oceanic crust
Eurasian plate
Indian plate
Very old rock, 2–2.5 billion years old
Reference point

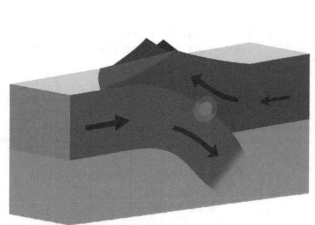

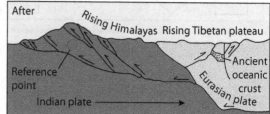

After
Rising Himalayas
Rising Tibetan plateau
Reference point
Ancient oceanic crust
Indian plate
Eurasian plate

Fig 3.14 Collision Plate Boundary

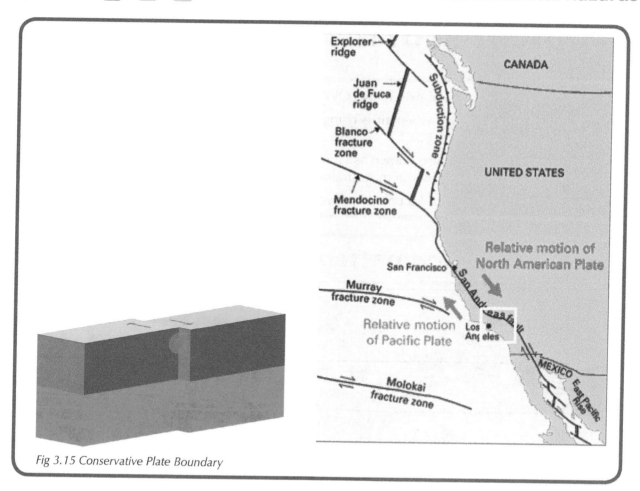

Fig 3.15 Conservative Plate Boundary

Convection currents

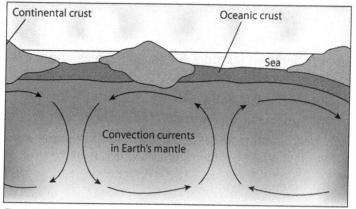

Fig 3.16 Convection Currents

- Plates move due to convection currents within the mantle.
- Heat from the Earth's core causes the mantle to rise up towards the crust where it is cooled and sinks back down to be heated again.
- The circular movements of magma cause plates to move very slowly.
- Very small movements of the plates can have dramatic consequences!

Tropical storms

A tropical storm is an area of extremely LOW atmospheric air pressure.

LOW air pressure is created when air is rising, causing LOW pressure on the Earth's surface.

Severe tropical storms are also known as Hurricanes (Atlantic), Tropical Cyclones (South East Asia), **Cyclones** (northern Australia) and Typhoons (western Pacific).

About 500 million people in 50 countries live in areas which could be affected.

The distribution pattern of tropical storms

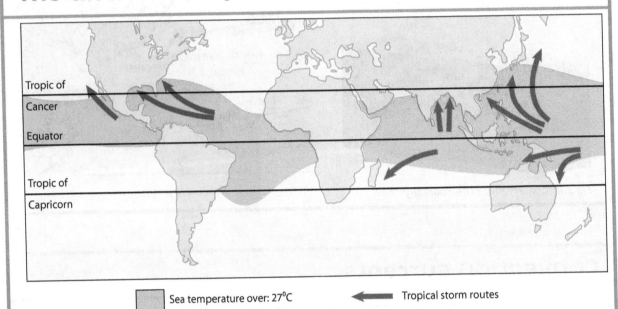

Fig 3.17 The distribution pattern of tropical storms

- Tropical storms occur between latitudes 5 and 30 degrees north and south of the equator.
- North of the Equator they occur in the Atlantic, eastern Pacific and the western Pacific.
- South of the equator they occur off Australia and in the Indian Ocean.
- They develop only over warm tropical oceans where sea temperatures exceed 27°C. centigrade and there is a considerable depth of warm water (60m).

Conditions necessary for a tropical storm to occur

1. Warm seas, surface temp 27°C, warm water to depth of 60m

2. Low air pressure (rising air)

3. Damp, moist air (60% humidity)

When these conditions are found, a tropical storm can occur.

Stages in the formation of a tropical storm

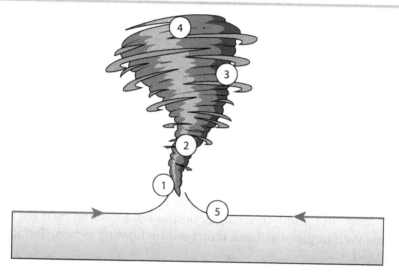

There are five stages in the formation of a tropical storm:

1. Warm damp air rises rapidly

2. Air begins to spiral because of the spinning of the earth (coriolis effect)

3. Warm air cools as it rises, condenses, forms clouds, brings rain

4. Air spreads out

5. Air rises and spreads so more damp air is sucked in to replace it.

Hurricane categories

The most severe hurricanes are Category 5 hurricanes. The intensity or strength of hurricanes and tropical storms elsewhere is measured on the Saffir-Simpson scale.

SAFFIR-SIMPSON HURRICANE SCALE

Tropical Storm

Winds 39–73 mph

Category 1 Hurricane — winds 74-95 mph (64–82 kt)

No real damage to buildings. Damage to unanchored mobile homes. Some damage to poorly constructed signs. Also, some coastal flooding and minor pier damage.

- Examples: Irene 1999 and Allison 1995

Category 2 Hurricane — winds 96–110 mph (83–95 kt)

Some damage to building roofs, doors and windows. Considerable damage to mobile homes. Flooding damages piers and small craft in unprotected moorings may break their moorings. Some trees blown down.

- Examples: Bonnie 1998, Georges(FL & LA) 1998 and Gloria 1985

Category 3 Hurricane — winds 111–130 mph (96–113 kt)

Some structural damage to small residences and utility buildings. Large trees blown down. Mobile homes and poorly built signs destroyed. Flooding near the coast destroys smaller structures with larger structures damaged by floating debris. Terrain may be flooded well inland.

- Examples: Keith 2000, Fran 1996, Opal 1995, Alicia 1983 and Betsy 1965

Category 4 Hurricane — winds 131–155 mph (114–135 kt)

More extensive curtainwall failures with some complete roof structure failure on small residences. Major erosion of beach areas. Terrain may be flooded well inland.

- Examples: Hugo 1989 and Donna 1960

Category 5 Hurricane — winds 156 mph and up (135+ kt)

Complete roof failure on many residences and industrial buildings. Some complete building failures with small utility buildings blown over or away. Flooding causes major damage to lower floors of all structures near the shoreline. Massive evacuation of residential areas may be required.

- Examples: Andrew(FL) 1992, Camille 1969 and Labor Day 1935

Fig 3.18 The Saffir-Simpson Scale

Weather associated with a tropical storm

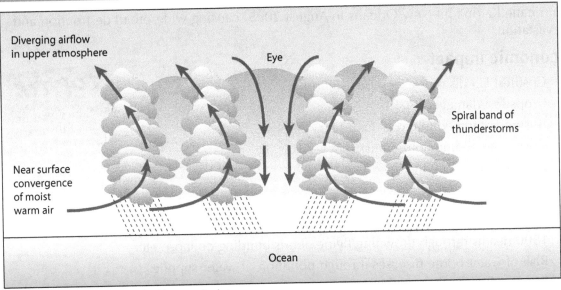

Fig 3.19 Weather associated with a tropical storm

Weather conditions with the passage of a tropical storm

Weather Element	The storm approaches	The storm worsens	The eye	The storm eases	The storm goes away
Air Pressure	Falls	Falls rapidly	Very low	Begins to rise	Rises
Temperature	Falls	Falls	Rises	Falls	Rises
Wind Speed	Increases	Increases to at least 60km/hr	Calm	Increases again	Decreases
Cloud Cover and Type	Cumulus clouds begin to form	Cumulonimbus clouds	Clear No cloud	Huge clouds build up	Clouds clear to give sunny spells
Precipitation amount and type	Starts to rain showers	Torrential rain (thunderstorms)	Dry	Torrential rain	Less rain Becomes dry

The impact of tropical storms

Hurricane Katrina hit New Orleans in August 2005, causing widespread destruction and devastation.

Economic impact

- Costliest US disaster ever – $81 billion
- Crops lost/damaged
- Businesses forced to close due to lack of power
- Cost of rebuilding homes/businesses, roads etc.
- Cost of emergency services
- Cost of insurance

TOP TIP

Make sure you have specific information related to your named Tropical Storm to include in your answer

Social impact

- 1500 deaths through drowning/flying debris/building collapse etc.
- Risk of water borne diseases through pollution of water supplies
- Food and water shortages
- Homelessness – more than 800,000 homes destroyed
- Many injured
- Crimes such as looting in downtown New Orleans
- Shortage of medical care/supplies
- Communications disrupted (telephone lines, roads etc.)

Environmental impact

- Trees uprooted
- Landslides
- Flooding
- Deposition of coastal/river material
- Water pollution

Limiting the impact of tropical storms

The effects of tropical storms can be reduced via the implementation of the following:

- Accurate forecasting of hurricanes using methods such as satellites (cloud)/weather stations/radiosonde balloons/weather planes
- Tracking the path of hurricanes using satellite and radar
- Warning systems through media
- Education of public/practice evacuations
- Implementation of disaster plans use of disaster supply kits

The role of aid agencies when a tropical storm/hurricane occurs

Short term aid from Developed Countries and charities includes:

- Search and rescue: military aid in form of helicopter lifts/rescue victims
- Food supplies and drinking water
- Blankets and tents to provide shelter
- Medicines and first aid supplies to treat injured persons

Long term aid:

- Money: Government funding for rebuilding houses/roads/businesses
- Building supplies (especially for rebuilding houses, schools, hospitals)
- Seeds and basic implements to get farmers back on track
- Medical supplies to re-stock hospitals
- Counselling and bereavement services

Quick Test

1. Describe the distribution pattern of tropical storms.
2. Explain, in detail, the ways in which aid can lessen the impact of a tropical storm. Your answer should refer to examples you have studied.

Earthquakes

Earthquakes

An earthquake is a sharp sudden movement between two plates of the Earth's crust. A huge amount of energy is released. The energy moves out in the form of shock waves (these are called **seismic waves**).

The point where the earth movement takes place is called the **focus** or origin. Shock waves are felt first and most strongly at the **epicentre**, the point on the earth's surface above the focus. Moving away from the epicentre, less damage occurs. In an earthquake, shock waves cause the most damage.

The magnitude of an earthquake is recorded by an instrument called a **seismometer**. It measures the height of the shock waves on the **Richter Scale**. The Richter Scale measures the amount of energy released by an earthquake on a logarithmic scale. Each point on the scale is 10 times greater than the one before.

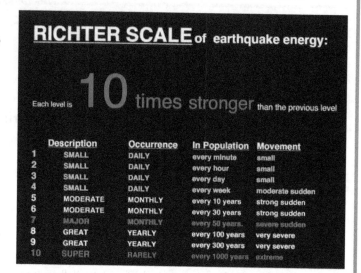

RICHTER SCALE of earthquake energy:

Each level is **10** times stronger than the previous level

	Description	Occurrence	In Population	Movement
1	SMALL	DAILY	every minute	small
2	SMALL	DAILY	every hour	small
3	SMALL	DAILY	every day	small
4	SMALL	DAILY	every week	moderate sudden
5	MODERATE	MONTHLY	every 10 years	strong sudden
6	MODERATE	MONTHLY	every 30 years	strong sudden
7	MAJOR	MONTHLY	every 50 years	severe sudden
8	GREAT	YEARLY	every 100 years	very severe
9	GREAT	YEARLY	every 300 years	very severe
10	SUPER	RARELY	every 1000 years	extreme

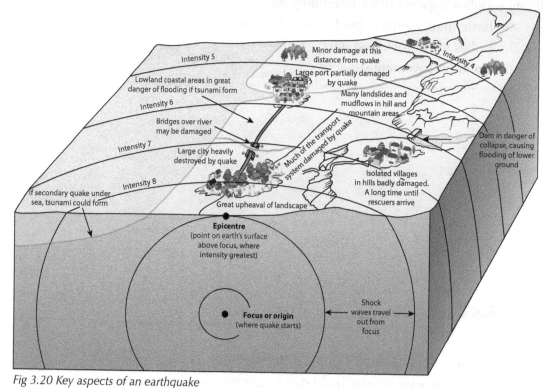

Fig 3.20 Key aspects of an earthquake

Global distribution of earthquakes

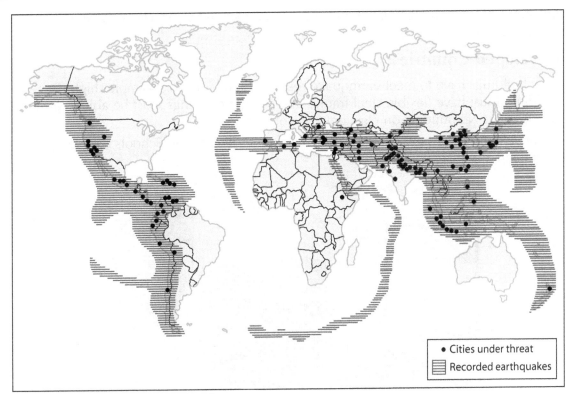

Fig 3.21 Global distribution of earthquakes

Earthquakes are concentrated in just a few parts of the world. They occur near crustal plate boundaries, where plates are **moving past or are colliding** with one another. They are particularly common around the edge of the Pacific ocean (e.g. California, Japan). Most occur under the seas because most plate boundaries are found there.

Methods used to cope with earthquakes

Earthquakes can never be prevented, but risk can be reduced using *The 3P's policy*: predict, protect, prepare.

In Developed Countries:

- New buildings have to meet earthquake-resistant standards, for example, high-rise buildings must have flexible steel frames and smaller buildings must be able to absorb shockwaves. No building on clay, only on solid rock.

- Earthquake drills (as per fire drills in the UK) are carried out in Schools e.g. 'Drop, Cover and Hold'

- Take cover under a desk or table
- Hold on to the desk/table leg so it stays on top of you
- Stay underneath until shaking stops

Fig 3.22 Drop, Cover and Hold

- Training of emergency services (particularly in terms of search & rescue)
- Families/households have to put in place an earthquake emergency plan and survival kit
- They have set up a TV, mobile phone messaging and radio warning system

In Developing Countries:

Unfortunately, developing countries do not have the finance to put any real structures or plans in place.

Therefore, the effects of earthquakes are usually much greater:

- Buildings in LEDCs are often poorly constructed and therefore more easily destroyed
- There is a lack of technology to help predict so people do not have any warning
- Poor search and rescue efforts due to a lack of training as there is no funding
- In some cases, developing countries will be more densely populated (e.g. Port au Prince in Haiti) so more people will be affected
- Developing countries rely heavily upon International Aid which takes longer to arrive

The impact of Earthquakes

CASE STUDY, HAITI – 12 JANUARY 2010, 21:53 UTC

Haiti fact file

Location – Caribbean

Capital City – Port au Prince

Number of inhabitants –4 million

Economy – poor economy; LEDC

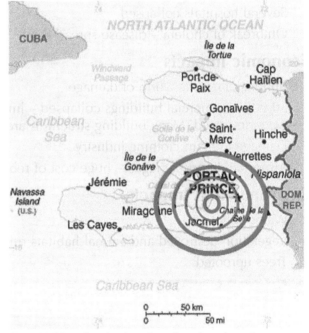

Fig 3.23 Map showing location of earthquake epicentre

Haiti earthquake – key facts

Magnitude 7.0

Tuesday, January 12, 2010 at 21:53:09 UTC

Epicentre 10 miles west of the capital of Haiti, Port-au-Prince

Conservative Plate Boundary – Movement of Caribbean and North American plates

Local population – over 3 million people immediately affected

What was the impact?

Social Impacts

- 220,000 estimated to have died
- 300,000+ injured
- 1.5 million homeless and living in camps
- Several hospitals collapsed
- Outbreak of cholera – disease spread easily

Economic Impacts

- Billions of dollars worth of damage
- 30,000 commercial buildings collapsed – impacting on local economy as businesses were destroyed. Many building structures are not earthquake proof
- Damage to main clothing industry
- Airport and port damaged – huge cost of rebuilding

Environmental Impacts

- Water pollution
- Vegetation destroyed and animal habitats ruined
- Trees uprooted

Who helped? AID after the earthquake

- Neighbouring Dominican Republic provided **emergency water** and **medical supplies** as well as heavy machinery to help with search and rescue underneath the rubble, but most people were left to dig through the rubble by hand.
- Emergency **rescue teams** arrived from a number of countries, e.g. Iceland.
- Medical teams began treating the injured – **temporary field hospitals** were set up by organisations like the International Committee of the Red Cross.
- **GIS** was used to provide satellite images and maps of the area, to assist aid organisations.
- People from around the world watched the news from Haiti on TV and through social networks. Many **pledged money** over their mobile phones.
- United Nations troops and police were sent to help distribute **aid** (water, food and shelter) and keep order.

Quick Test

For a named earthquake you have studied explain, in detail, the impacts on people and the landscape.

Volcanoes

Volcanoes

A volcano is an opening or vent in the Earth's crust. Magma (molten rock) from deep inside the earth's mantle forces its way to the surface. The magma may appear flows of molten lava, volcanic bombs, fragments of rock or simply as ash and dust.

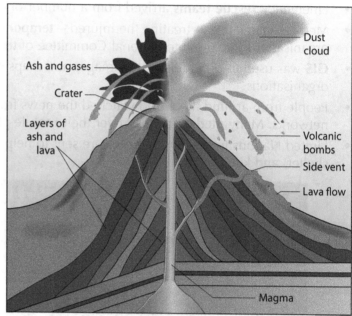

Dust cloud

Ash and gases

Crater

Layers of ash and lava

Volcanic bombs

Side vent

Lava flow

Magma

Fig 3.24 Volcano cross-section

Classifying volcanoes

Extinct: Volcanoes that haven't erupted since recorded history and show no signs of ever erupting, e.g. Arthur's Seat, Edinburgh

Dormant: Volcanoes that have not erupted in historic time but probably will erupt again, e.g. Mauna Kei, Hawaii

Active: Volcanoes that have erupted in recent times, e.g. Mt St.Helen's in USA or Eyjafjallajokull in Iceland

The distribution pattern of volcanoes

Most volcanic activity is found along plate boundaries but some occur as hotspots where the crust is relatively thin.

Many volcanoes are located around the Pacific Ocean or Ring of Fire. They occur on the west coast of South America where the Nazca plate meets the South American plate, and is forced underneath it, and along the mountain ranges of the Andes in South America and the Rockies in North America.

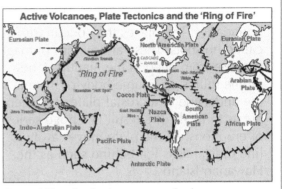

Fig 3.25 The distribution pattern of volcanoes

Why volcanoes occur at Destructive Plate Boundaries

- Plates move towards each other
- Oceanic crust is <u>more dense</u> so it is subducted (forced down)
- The crust melts in the mantle creating more magma
- Pressure builds and is released when magma is forced through a vent or opening
- This is a volcanic eruption, which can lead to the formation of mountains e.g. Andes

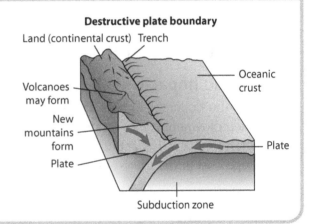

Why volcanoes occur at Constructive Plate Boundaries

- Two plates move away from each other
- Plates move due to convection currents in the mantle
- This creates a vent which allows magma to be forced through the crust to the surface
- Magma on the surface is known as lava which solidifies into rock creating new land
- E.g. Mid Atlantic Ridge which runs through Iceland

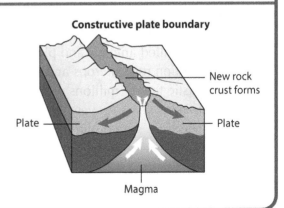

Hotspots

Some volcanoes are NOT found on plate boundaries. The places they occur are called hotspots. These are plumes of magma that rise from deep in the mantle. A new plume builds up a large body of magma on the underside of a crustal plate, and then bursts through. The result is a huge series of eruptions. As the plate moves away a line of islands is left behind such as the Hawaiian Islands.

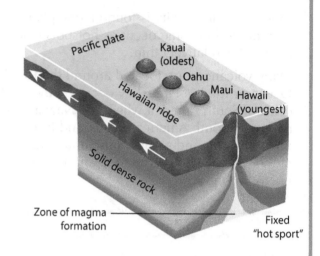

The impact of Volcanoes

CASE STUDY, Mt St Helen's – 18 MAY 1980, 08.32

Economic Impacts
- $2.74bn dollars' worth of damage
- Unemployment = loss of income
- Loss of crops = impact on agricultural industry
- Huge cost to the tourist industry and the logging industry
- Increase in national debt

Social Impacts
- 57 people dead
- 543 people injured
- 8,000 homeless
- Telephone and electricity lines were damaged so people were 'cut off'
- Loss of jobs in the tourist industry

Environmental or Physical Impacts
- Mudflow killed animals and plants, choked rivers killing all fish
- Ash eruption ruined crops and caused air pollution
- Pyroclastic flow = millions of trees uprooted
- It took many years for vegetation in the area to start to grow back
- Changed the shape of the mountain (large crater in the side)

Limiting the impact of Volcanoes

The effects of volcanic eruptions can be reduced via the implementation of the following:

- Monitor seismic activity
- Watch for changes in gas levels/sample gas emissions
- Patterns of previous eruptions
- For Mt St Helen's – growth of 'bulge'
- Unusual animal/wildlife behaviour
- Implement an exclusion zone to restrict access near the volcano

How effective are the predictions?

- Scientists can predict where, but not when – Mt St Helen's anticipated but not for another few weeks
- An exclusion zone was created but it was not extensive enough as 90% of people were killed outside the exclusion zone
- Government preparation was taken but the blast was simply underestimated
- Some people took risks and did not listen to advice

The positive impacts of volcanoes

Not all impacts of volcanic eruptions are disastrous. There can be many positive outcomes, including:

- Volcanic ash creates fertile soils which is good for growing tomatoes, grapes and olives.
- There are job opportunities related to the volcano e.g. guiding, and working in cafes and hotels or selling souvenirs e.g. jewellery, postcards. This benefits the local economy.
- The scenery is stunning/attractive.
- Volcanic eruptions deposit minerals on the surface of the earth so they can be used by humans, e.g. sulphur and copper.
- The climate is warm and pleasant all year round in the Bay of Naples and on Sicily.
- In Iceland, geothermal energy is used for heating and cheap hot water.

Quick Test

1. Describe the distribution pattern of volcanoes.
2. For the Icelandic eruption, or any other volcanic eruption you have studied, describe the effects of the eruption on the people **and** the surrounding landscape.

Trade and globalisation

Trade is the exchange of goods and services between countries. Trade is necessary as no country in the world is self-sufficient (i.e. supplies everything it needs). So countries have to:

- **Import** – buy goods they need from other countries.
- **Export** – sell goods they produce to other countries.
- **Have a positive balance of trade** – the difference in value between imports and exports.

If a country buys more than it sells, that country will have a trade deficit. If a country sells more than it buys it will have a trade surplus. If the imports equal exports then a country is said to have a balance of trade. World trade patterns are very important as they determine a country's economy and hence its level of development.

- Developed countries rely on the import of raw materials or primary goods, which they then process and turn into manufactured goods for export.

- Developing countries are still the main producers of primary goods, which they export, but they have to often import manufactured goods. The price a developing country is paid for a raw material or primary good is a fraction of the price paid for the manufactured good it is made into.

- Developed countries often have a trade surplus. Developing countries often have a trade deficit. However, the UK actually has a trade deficit and many countries in both the developed and the developing world are not really paying their way in terms of production but are instead mortgaging future production. The UK is currently paying £1 billion per week in interest on its debts.

However, developed and developing countries are interdependent of each other.

DID YOU KNOW?

Developing countries have only a small share of world trade and prices for goods are determined by the developed countries.

Newly industrialised countries

Newly industrialised countries (NICs) sit 'between' developing and developed status. They have growing economies and are experiencing rapid industrialisation, for example China, Brazil, India, Mexico, South Africa, Turkey and Malaysia.

EXAM TIP

Make sure you can identify the exports of at least two of these countries. An explanation of why these countries are changing is shown opposite.

The Clark-Fisher Model

This model shows that as a country's economy grows the relative importance of its three sectors (primary, secondary and tertiary) change.

As an economy grows, manufacturing industries grow and develop. Factories spring up and tertiary industries start to grow as the population becomes more urban.

Developing countries have a large amount of primary industries. There are very few jobs in manufacturing (secondary) or tertiary services (banking, insurance, investment).

In this stage, the tertiary sector becomes more important than the secondary sector. More people living in towns and cities means more demand for tertiary or service types of industry. The quaternary (high-tech/research) sector starts to grow.

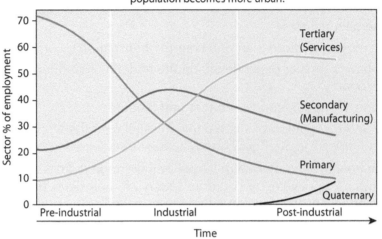

Examples: Zambia, Ghana, Kenya, Egypt, Algeria.

Examples: Malaysia, Turkey, Mexico, Brazil, India, South Africa.

Examples: UK, France, Germany, USA, Japan.

Fig 3.26 The Clark-Fisher model

Other reasons for their success are:

- access to raw materials
- cheap and available labour
- highly skilled workforce
- tax incentives

Quick Test

1. Explain the difference between imports and exports.
2. What is meant by the 'balance of trade'?
3. Why do developing countries often have a trade deficit?
4. Explain why developed and developing countries are 'interdependent'.

Trade and globalisation

The world has become increasingly interconnected as a result of increased trade, telecommunications, the Internet and the exchange of cultures. This is referred to as 'globalisation'. Some of the world's biggest companies are no longer national but international (operating in many countries around the world).

Globalisation has meant:

- increased international trade
- companies operating in more than one country (transnational corporations or TNCs)
- all countries have a greater dependence on the trade fluctuations (up and downs) of the world's economy
- freer movement of capital, goods, services and labour
- the presence and recognition of some companies all over the world, even in developing countries e.g. Starbucks and Macdonald's

Many developing countries are in debt because of their trade deficits and they are actually paying money to developed countries to pay off their debt. This makes developed countries richer and means the developing countries have less money to invest in manufacturing, education and health care.

Low labour costs in developing countries

Improvements in transport

Labour availability and skills

Reasons for globalisation

Freedom of trade

Better communication

Trading alliances

Trading alliances

Some countries group together for economic reasons. Examples of trading alliances are the EU, NAFTA (North American Free Trade Association) and OPEC (Organisation of Petroleum Exporting Countries). Member states within a trading alliance gain certain benefits:

- No tariffs (import taxes) on goods traded between member states
- Free movement of labour
- All countries contribute money to the alliance and the resources of the member countries can be more evenly distributed
- Control may be put on the price of a raw material, for example oil by the OPEC alliance

Despite these alliances, developed countries still dominate world trade. In 2008 the five largest importers and exporters were Germany, USA, France, Japan and UK. These have now been joined by China, which has pushed the UK into sixth place. These countries account for 81% of all trade.

Transnational corporations

A transnational corporation (TNC) is also called a multinational company. These TNCs often have headquarters in a developed country and increasingly factories in developing countries. Some examples of TNCs are Coca-Cola, Microsoft, Dell, Honda, Sony, Nestlé, Royal Dutch Shell and Vodafone. In the past, manufacturers and oil companies have been the largest corporations, but telecommunications are beginning to catch up. 75% of world trade is carried out by these corporations and their benefits to the developing world are often disputed. This table lists the advantages and disadvantages of TNCs to a developing country.

Advantages	Disadvantages
TNCs bring work and employ local labour.	The presence of TNCs encourages the local young people into factories rather than into education.
TNCs often pay higher than the national average.	Profits of the corporations go back to developed countries.
TNCs bring improvements to infrastructure e.g. roads and rail.	The number of local employees is actually quite low.
TNCs bring in other foreign investment.	The work of TNCs often has negative environmental impacts, for example oil spills in Nigeria.

Fig 3.27 Advantages and disadvantages of Transnational corporations

Quick Test

1. Give three reasons for globalisation.
2. Identify two benefits of forming a trading alliance.
3. Give two negative impacts of a TNC (Transnational Corporation) in a developing country.

The developing world fights back

Developing countries lose out on the value of a product by exporting them in their raw state rather than processing them. One solution to increase earnings is to tax these exported raw materials. This may make them less attractive on world markets but it would significantly increase the country's income.

Case study: Kenyan leather trade

Africa produces 15% of the world's cow hides and 25% of the world's goat and sheep skins. Yet its global share of leather production is only 2% because most of it is processed elsewhere. Here is how Kenya tried to address this:

The 1990s Kenyan Government reduced the taxes on leather goods coming into the country.

↓

This lead to a crisis as the Kenyan leather trade could not compete with the cheap imports.

↓

The Kenyan Government lost millions of pounds in revenue.

↓

In 2006, Kenya launched its Vision 2030 policy and an export tax was put on raw hides and skins.

↓

This meant that it was better for the hides and skins to be processed into leather in Kenya. Now between 96 and 98% of hides and skins were processed in Kenyan tanneries. This lead to higher export earnings, more jobs and better wages.

Fig 3.28 A Kenyan woman creates a bag from goat skin

Fair trade

Fair trade guarantees good prices for produce – usually raw materials such as bananas, coffee and cocoa – and this helps to create stability and benefits the whole community.

- Fair trade pays a guaranteed minimum price to the grower to protect them from world price fluctuations.

- At least 20% of the profit goes to the grower in an effort to cut out the middleman (the importer, e.g. UK supermarkets).

Fair trade products do cost more but shoppers need to understand that long-term this will reduce inequalities in world trade and help producers in developing countries.

'Fair Trade is about making sure people get their fair share of the pie. Nobody wants to buy something that was made by exploiting somebody else.'

Jerry Greenfield from Ben & Jerry's Ice Cream, 2010

Fig 3.29 Coffee is a popular fairtrade product

Quick Test

1. What is fair trade?

2. What percentage of the profit for the produce, e.g. bananas, goes to the grower/producer?

What is global tourism?

Global tourism is a multibillion pound industry that is one of the leading sources of income in the world and has a huge impact on the economic development of people, areas and individual countries.

In 2016 international tourism receipts earned by destinations around the world totaled US$ 1.235 trillion. International tourism accounts for 7% of the world's exports in goods and service and has grown faster than world trade for the past five years. Its growth has remained uninterrupted for decades, showing the sector's resilience. Tourism has a positive economic impact on both the developed and developing parts of the world. Countries see tourism as an important aspect of future development, creating jobs, increasing export revenues and improving infrastructure. Worldwide, tourism is responsible for 1 in 10 jobs.

World tourism in figures (2016)

Leading tourism earners 2016		International tourist arrivals	
Country	International tourism receipts ($US billion)	Country	International tourist arrivals (million)
USA	205.9	France	82.6
Spain	60.3	USA	75.6
Thailand	49.9	Spain	75.6
China	44.4	China	59.3
France	42.5	Italy	52.4
Italy	40.2	United Kingdom	35.8
United Kingdom	39.6	Germany	35.6
Germany	37.4	Mexico	35.0
Hong Kong (China)	32.9	Thailand	32.6
Australia	32.4		

Fig 3.30 Leading tourist earners and fastest growing tourist economies, 2016

The impact of tourism

Although it plays an important part in economic development, tourism also has an effect on people and the environment. The key word with tourism is **sustainability**. If tourism is not managed effectively it may cause problems that limit its economic benefits and may even kill off tourism as the attractiveness of a tourist area declines through the impact of too many people.

What is the Butler Model of Tourism Growth?

The Butler Model of Tourism Growth uses the experience of tourism development in, for example, Europe since the 1960s, to identify how a tourist economy will develop and decline or re-invent itself in a sustainable way.

Stage 3 Success: The tourist industry develops to the point where it reaches its full potential with all its facilities working to full capacity. Little of the original natural environment is left and the lives of people in the area will have changed significantly along with traditional towns, villages and traditional lifestyles. The area loses its unique charms and becomes rather like the rest of the world (globalisation).

Stage 2 Growth/Development: Initial interest in the area grows and spreads through the media and by word of mouth. Special tourist infrastructure such as hotels, roads, airports and transport termini are built along with improvements to water and sewage provision. This all provides jobs and money for the economy.

Stage 1 Discovery: A new tourist area is identified by a small number of visitors based for example on its wildlife interest, its historic attractions, its potential for outdoor leisure pursuits, its special culture or its idyllic tourist attractions.

Fig 3.31 The Butler Model

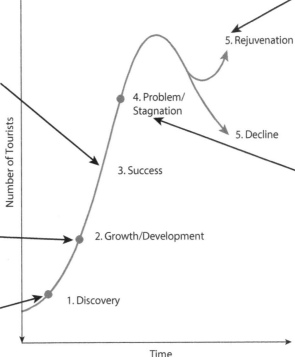

Stage 5 Rejuvenation or Decline? The tourist area has an option here to deal with the reasons for decline and invest money to re-brand and re-develop the area or continue with declining tourist income and falling employment.

Stage 4 Problem/Stagnation: People become bored with the area or the area becomes spoiled and visitors move on to newer developments elsewhere which are less spoiled by tourism. The tourist area goes into economic decline and the tourist facilities become jaded and in need of an upgrade.

EXAM TIP

In this section of the course it is important to choose case studies of tourist development and eco-tourism. Use the Internet to research a particular part of the world you are interested in. Why not choose a place you have visited on holiday and work out how much impact you and your family had on the economy and environment of the area? Questions like those in the Quick Test can be used (along with others) to focus your study.

Quick Test

1. Why is tourism important on a global scale?
2. How can success in developing a tourist area lead to problems?
3. What would be the effect of stagnation on a tourist area?

What impact does tourism have on an area?

Many people complain about the impact of tourism on an area and how it might spoil the very places that visitors come to see and enjoy. Eco-tourism might be defined as being all about people visiting environmentally sensitive areas, but if the visitors create problems for landscape, cultural features such as historic sites and wildlife, then tourism is not sustainable. Eco-tourism should be about visitors having little or no impact on the original environment and about management systems being put in place that achieve the vital tourist income but protect the environmentally sensitive area.

Managing tourism and related problems

Tourist activity	Problem caused	Management strategy
Large numbers of tourists visit an environmentally sensitive area	• Wildlife is disturbed and moves away • Trails and paths are developed • Accommodation puts demands on water and local fuel resources	• Designate national parks and nature reserves • Restrict access to reduced areas • Use renewable energy sources rather than local wood
Beach and coastal tourism	• Sprawling urban development along the coast • Sewage disposal and environmental pollution • Demands on water resources	• Develop planning guidelines to allow appropriate development • Build sewage disposal capacity • Conserve water supplies by recycling waste water • Protect sensitive coastal environments by creating marine reserves or restricting access to reefs for divers
Visiting historic sites	• Visitor numbers cause erosion on ancient buildings by footfall • Entering old buildings may damage wall paintings and artefacts	• Restrict access to fewer people • Build replicas of the original site such as Tutankhamen's tomb or the Lascaux caves and their paintings
Safari and wildlife experiences	• Visitors disturb wildlife and local people disrupting their 'natural' way of life • Animals are habituated to people being present and become a danger to locals and visitors	• Develop low impact visit strategies that keep a distance between people and the target wildlife • Restrict access maintaining off limit parts of reserves • Reduce hunting and camping in the field

Ski development	• Visitors generate demand for ski facilities and uplift (this can spoil sensitive mountain environments) • High demand for water and sewage infrastructure creates supply and disposal problems	• Develop environment guidelines for ski development • Conduct environmental impact studies and ecological surveys • Assess new avalanche risks caused by tourist development
Waste management	• Visitors produce large amounts of rubbish and waste in environmentally sensitive areas that may not be disposed of appropriately	• Provide public education and guidelines that reduce packaging • Recycle waste and remove it to proper facilities • Reduce reuse and recycle campaigns

Fig 3.32 Managing tourism

What is mass tourism?

Some areas like the Spanish Costas, Florida, and major ski resorts attract large numbers of tourist visitors. Cheap package holidays are available and people can stay in apartments and large hotel complexes. Local business people and large leisure companies provide facilities for tourists to use and enjoy. Beaches and attractions can become very busy, local services expand and involve heavy traffic in the tourist season. Sometimes large visitor numbers create environmental pressure or places become too crowded and begin to discourage tourists who look to other newer destinations for holidays.

Tourism continues to grow in world terms as people begin to earn more money from industrial jobs and have more leisure time. As countries like China and India see growth in their economies they too are developing tourist activity and seeing more and more people travelling further afield. Trends in tourism seen for many years in developed countries are now being mirrored in the developing economies.

Fig 3.33 A crowded beach in Costa del Sol, Spain

Quick Test

Why are some areas particularly attractive to tourists? (Use specific examples such as the Alpine ski areas, the Caribbean Islands, Africa's national parks (safaris), rainforests, spectacular scenery areas, Florida, Spain's tourist coasts and holiday islands.)

What is eco-tourism?

Eco-tourism is a rapidly expanding market. As people look for new holiday ideas they may journey to increasingly exotic locations encouraged by such things as greater accessibility from cheap air fares etc. Wider knowledge of the world through media sources such as TV programmes creates a demand for access to environmentally sensitive areas so that visitor numbers once associated with popular mass tourist destinations may be finding their way to important wildlife and cultural centres. This may create new problems like those already seen in popular tourist hot spots.

Eco-tourist destinations that are under threat

- Safari holidays to Africa visit environmentally sensitive areas and may not be able to manage large numbers of visitors in what was previously an exclusive activity. New hotels encroach on national parks and animal habitats. Water supply and sewage management may cause issues.
- Whale watching trips can cause disturbance to wildlife if they are un-regulated. Scuba diving can have an impact on reefs such as the Great Barrier Reef of Australia or the Red Sea. Air boat tours in Florida cause disturbance.
- Trekking in the Himalayas puts demands on local water or wood supplies and therefore has an impact on forests in Nepal. This leads to increased pressure on the land and to conflict between tourists and farmers. Loss of forest creates major flood and soil erosion problems.
- Visits to cultural sites such as Machu Picchu or the Pyramids in Egypt turns these into tourist honey pots and eventually begins to damage the cultural heritage. People visit tombs and historic sites in large numbers and excessive footfall and even the problems caused by the effects of breathing out carbon dioxide onto the walls can affect monuments and heritage sites.

Conserving heritage sites and environmentally sensitive areas

The Lascaux Caves in France are a World Heritage Site famous for their cave paintings. Visitors are no longer allowed to visit the caves because of the damage that was caused to the priceless Stone Age art works. Instead 'New Lascaux' was built as an exact replica so that visitors can still get an impression of the original site.

Fig 3.34 Mortuary Temple of Hatshepsut, near the Valley of the Kings, in Luxor, Egypt

At the Valley of the Kings in Egypt four hundred or more tourists visit the tombs each day, each leaving behind an ounce of moisture – about a third of a teacup – from their breath. Bright lights are needed so that visitors can see, and these raise the temperature in the tombs; these and other factors are slowly destroying the tombs and the paintings inside. Scientists who have identified this problem have developed a plan to build new visitor centres with simulations like those built at Lascaux, along with roads and flood control systems to protect this World Heritage Site from further damage.

Endangered species like orangutans, chimpanzees or mountain gorillas are now mostly found in nature reserves. Visitors may contribute to their eventual survival by providing conservation income but may equally contribute to their extinction by disturbing their increasingly threatened habitat.

Eco-tourism can be an important way of preserving the environment and ecology of an area but it needs to be managed in such a way that it is sustainable and the impact of visitors is minimised.

DID YOU KNOW?

You can calculate your tourism carbon footprint by doing some research using the Internet. This might let you plan a holiday that makes a minimum environmental impact. One search delivered the following site: www.carbonzero.co.nz.

Quick Test

1.   What is eco-tourism?
2.   Why do tourist developments in environmentally sensitive areas have to be carefully planned?
3.   Why do theme parks and visitor attractions like the Disney Parks put pressure on the environment?

Where and why does tourism develop around the globe?

City heritage sites

Cities with major historic, cultural and entertainment functions.

Fig 3.35 New york city

Reasons for location

Barcelona – Cultural heritage, cruise liner port, architecture (Gaudi), art (Picasso Museum), lively city atmosphere at Las Ramblas, beaches and hotels, football and Olympic heritage.

Paris – Historic city centre, classic churches (Notre Dame), museums and art galleries (Louvre), Monuments (Eiffel Tower, Arc de Triomphe), night life, restaurants and bars.

Prague – Historic city (castle/palace), lively city centre, restaurants and bars, short city break weekend destination.

New York – Modern city architecture, museums and galleries, tourist hotspots (Central Park), film locations, theatre and musical district (Broadway), tourist attractions (Statue of Liberty, Time Square, Wall Street), shopping, bars and restaurants.

Environmental impact and management issues

- Cities become overcrowded, traffic conjestion, honeypots and main attractions become very busy.
- Historic buildings lose their charm – other users such as churchgoers are affected.
- Sites damaged by footfall and overuse.
- Weekend/short breaks may involve anti-social behaviour such as stag and hen parties.

National park and scenic areas

Areas with particularly attractive scenery and leisure facilities.

Reasons for location

Lake District National Park, UK – Lakeland scenery, associations with poets and writers (Wordsworth, Beatrix Potter), camping, outdoor pursuits and motor touring.

Yellowstone National Park, USA – Volcanic scenery, geysers (Old Faithful), wildlife interest (bears, wolves and bald eagles).

Florida Everglades National Park, USA – Wetland heritage area, wildlife (birds and alligators), near to Miami, Orlando, and other Florida holiday destinations (Disney, Kennedy Space Centre), Florida Keys marine coastal nature reserves (John Pennekamp Reserve).

Environmental impact and management issues

- Range of visitor pressures on popular honeypot sites like those mentioned.
- Disturbance of wildlife and nature.
- Traffic congestion.
- Florida's popularity as a mass tourist destination causes urban sprawl, water management and pollution problems in the Everglades and similar wetland environments.
- Large numbers of visitors move on to other local sites beyond the Orlando Disney complex.
- Natural heritage under threat from development.

Ski resort development

Reasons for location

Cairngorms and Grampian Ski areas, Aviemore, European Alps, American Rockies – Skiing and winter sports are developed based on snow resources in the ski season. Tourist infrastucture is built on a large scale. Transport and uplift capacity is increased to serve tourists and visitors. Olympic and World Cup events bring in extra visitors and facilities.

Environmental impact and management issues

- Impact on wildlife and nature in scenic areas.
- Large tourist complexes change the nature of remote mountain districts.
- Forests areas are cleared.
- Demand is put on local land and water resources.

Quick Test

1. Why does building in wetlands provide environmental challenges?
2. Why do major events like the World Cup or the Olympic games cause concerns about their environmental impact?
3. What benefits do tourist developments bring to an area?

The distribution of world diseases: AIDS – a case study of a world disease

AIDS is an incurable disease caused by the Human Immunodeficiency Virus (HIV). It is a global disease but 70% of all AIDS/HIV cases are in Africa, and the distribution is very uneven.

Causes of AIDS

- HIV/AIDS is spread in three main ways:
  - Sharing a needle with an infected person.
  - Having unprotected sex with an infected person.
  - Babies drinking the breast milk of an infected mother.
- HIV/AIDS can also be spread by infected blood transfusions.
- A lack of education accelerates the spread of the disease as does a lack of contraception.
- Poverty is a key factor.

Economic effects/impact of AIDS

Developing countries
- Inability of workforce to deliver in their job – high absence rates through illness impacts on production levels.
- As adults become ill, children are forced to stop their education to look after their parents and try to earn money. The result is that the next generation of African adults will be less educated, less wealthy and possibly less healthy.
- Resources are diverted away from other projects into health care and particularly into the treatment of AIDS, which means less money and less health care for people with other diseases.

Developed countries
- People die younger so they have fewer years in the economically active bracket of population composition.
- As people become ill, they cannot work, so the cycle of ill health comes into effect, impacting on the economy.
- There will be fewer taxpayers so the country produces less wealth.
- → These are all also applicable to developing countries.

Social effects/impact of AIDS

- Stigma associated with HIV/AIDS can lead to mental health problems.
- Very large number of orphans = family unit breakdown.
- Negative effect on educaton = possible rise in illiteracy rates and uneducated nation.

Strategies to reduce the spread of AIDS

Developing countries
There is no cure for AIDS but the following can help reduce the impact:
- Health education programs
- Free testing for HIV
- Distribution of free condoms
- Reduce poverty
- Reduction in malnutrition

Developed countries
- Widespread availability of ARV (antiretroviral) drugs.
- Drugs to stop the disease passing from pregnant mothers to their babies.

> **EXAM TIP**
> Link the impact to the development of a case study country.

How effective are the controls for AIDS

The controls are not very effective, for the following reasons:
- There are not enough drugs or medication for everyone. Only 5% of babies who need to be treated actually receive the drugs.
- Poverty – countries cannot afford to buy the drugs and administer them.
- There are not enough trained staff to administer treatments.
- Many people are in denial or ignorant of the spread of disease.
- High illiteracy rates limit the success of education programs.

To some extent the controls work. Drugs allow children to remain well enough to receive an education (but they can struggle to keep up). Life expectancy increases, which allows family to remain together longer and reduces the number of children orphaned.

Quick Test

1. Explain the causes of AIDS
2. What are the consequences of the disease on the economy of a country?
3. How can the impact of AIDS be reduced?

Managing disease in developing countries

Diseases such as cholera, malaria, kwashiorkor and pneumonia are more prevalent in developing countries. You must know the causes, effects and strategies to manage **one disease** in detail, from the list above.

Case study – malaria

Malaria is a disease that is *endemic* in many countries – this means it is always present. Those who survive develop a limited measure of immunity in adulthood. They can however be re-infected, suffer from recurrent and violent bouts of fever and are severely anaemic.

Malaria is an infectious disease and is found mainly in the world's poorest tropical areas, such as Africa, South America and South-East Asia but apart from hot and cold deserts and high altitudes, mosquitoes have a worldwide distribution.

How malaria spreads

Malaria is a mosquito-borne, potentially fatal, blood disease. It is caused by a parasite that is transmitted to human and animal hosts by the *Anopheles* mosquito.

1. Infected mosquito bites a human.

2. The parasite is in the liver within 30 minutes.

3. The parasite multiplies rapidly in the liver. Parasites can lie dormant in the liver for years before being activated.

4. The parasite enters the blood stream and attaches to red blood cells; it multiplies further.

5. Infected red blood cells burst, infecting other blood cells.

6. The cycles repeats, reducing the oxygen in the body and causing fever. The cycle corresponds to the fever and chills experienced by someone who is sick with malaria.

7. After it has been released a dormant version of malaria stays in the blood stream. If an uninfected mosquito bites this person it will ingest the parasite and go on to infect someone else.

The impact of malaria

Malaria holds back development. In Thailand it costs the people nine times their daily income on drug treatments to control the disease. A bout of illness requires at least five days off work. The economy suffers as a result. Therefore there is not enough money to spend on education, which is crucial to the development of any country.

Methods used to control the spread of malaria

ABCDE!

A – be AWARE of the risk

B – avoid mosquito BITES

C – take your CHEMOPROPHYLAXIS (medication to help prevent malaria)

D – DIAGNOSE quickly

E – if more than 24 hours from medical attention, take a course of EMERGENCY stand-by treatment.

Other controls include work done by charities such as the Roll Back Malaria Campaign and the Bill & Melinda Gates Foundation. They support research and development for more effective treatment and spend time educating local people about what they can do. They believe that prevention is better than cure.

Quick Test

Using the Internet, research how effective these measures are and complete the table.

Preventative method	Effectiveness
Bed nets	
Window and door screens	
Insecticides	
Draining marshes and swamps (where mosquitoes easily breed)	
The latest anti-malarial drug artemisinin	A potential treatment: so far there has been little evidence of the parasite developing resistance to the drug, which is very positive, but there were some concerns about resistance in Indonesia a couple of years ago, so only time will tell.

Managing disease in developed countries

Diseases such as heart disease, cancer and asthma are more prevalent in developed countries. You must know the causes, effects and strategies to manage one of these diseases in detail.

Case study – heart disease

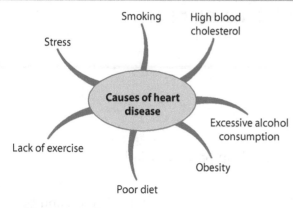

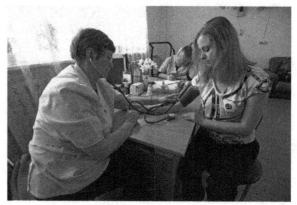

Fig 3.36 Doctor taking patient's blood pressure

Effects and consequences

- Inability to work.
- Lower life expectancy.
- Increased medical/health costs – operations, re-enforced beds for patients with obesity.

Strategies to manage the spread of heart disease

- Government policies in the UK, e.g. '5 A Day' the smoking ban and raising the age for purchasing cigarettes.
- More advanced medical care allows early diagnosis – prevention is better than cure.
- Advertising campaigns that encourage people to take more exercise, which helps prevent heart disease.
- Medical intervention – new drugs, clot removal and strengthening the arteries help control the disease.

Fig 3.37 People are encouraged to eat 5 different pieces of fruit and veg a day

Control methods

Many of the controls are unsuccessful:

- Despite health warnings, people continue to smoke and drink alcohol at unhealthy levels.

- People are not taking enough exercise and obesity levels are at their highest in the UK, so advertising campaigns do not appear to be working.

- Technology and gaming are the most popular pastimes for young people, which means they are not taking enough regular exercise and are key candidates for future heart disease.

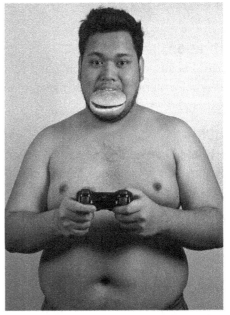

Fig 3.38 Despite warnings, poor diet and lack of exercise is common

Linking development and health

Question: How can levels of development help to reduce global health issues?

Answer: Increased development = better economy = more money available for health resources, better education, increased sanitation. This will help reduce disease and enable health care for illnesses.

Quick Test

1. For AIDS, heart disease, or malaria, what are the causes?
2. For a disease you have studied, what are the main consequences?
3. How effective are the controls?

Glossary

Abrasion
Erosion caused by the rubbing and scouring effect of material carried by rivers, glaciers, waves and wind.

AIDS
A disease of the human immune system caused by infection with human immunodeficiency virus (HIV).

Air Masses
Stationary air that takes on the characteristics of the region it is above.

Air Stream
Movement of air from these air masses.

Altitude
The height of the Earth's surface.

Amazonia
A large area of Brazil centred on the Amazon River and its tributaries. It is also called the Amazon River Basin.

Arête
A narrow, knife-edged ridge formed by glacial erosion when two adjacent corries erode back to back.

ARV drugs
Antiretroviral drugs used in the treatment of HIV infection.

Aspect
Direction in which a slope, building or settlement faces.

Brown field site
Land previously used for industrial purposes.

Butler Model
A model of tourism development over time showing the rise, development, decline and rejuvenation of tourism in an area.

Choropleth map
A thematic map in which areas are shaded or patterned in proportion to the measurement of the statistical variable being displayed on the map, such as population density or per-capita income.

Conurbation
A conurbation is formed when cities and towns expand sufficiently that their urban areas join up with each other.

Corrie
An armchair-shaped hollow caused by glacial erosion.

Deforestation
The cutting down of trees and clearing of forests.

Deposition
The process whereby eroded material is left behind on the landscape by such things as glaciers, rivers, the wind or the sea.

El Nino
A band of anomalously warm ocean water temperatures that occasionally develops off the western coast of South America and can cause climatic changes across the Pacific Ocean.

Endemic diseases
When a **disease** is prevalent in an area over long periods of time, it is considered to be **endemic** in that area.

Enhanced greenhouse effect
An increase in the natural process of the greenhouse effect, brought about by human activities like **burning fossil fuels** and **deforestation**, whereby greenhouse gases such as carbon dioxide, methane, chlorofluorocarbons and nitrous oxide are being released into the atmosphere at a far greater rate than would occur through natural processes and thus their concentrations are increasing.

Erosion
The wearing away of the landscape by river, glacial and marine processes such as abrasion, hydraulic action and attrition.

Evapotranspiraton
Loss of water from plants and the Earth's surface.

Favelas
A Brazilian term for an informal shanty-type settlement.

Freeze-thaw weathering
A process of physical weathering by which rock disintegrates, caused by water in cracks repeatedly freezing and thawing.

Gentrification
The process of renewing and rebuilding an area due to the influx of more affluent residents.

Glacier
A river of ice.

Global warming
An increase in the average temperature of Earth's atmosphere and oceans.

Green belt
An area of land around a large urban area where development is severely restricted.

Greenhouse gases
Gases in the atmosphere that absorb and emit radiation, e.g. carbon dioxide, water vapour, methane.

Hanging valley
A tributary valley left high above the main valley as its glacier was unable to erode as deeply as the glacier in the main valley.

Human development index
A composite statistic of life expectancy, education, and income indices used to rank countries according to their level of development.

Impermeable
A property, for example in rock layers, that does not allow water to pass through the structure.

Permeable structures allow the passage of water. Porous rocks have many interlinking air spaces that soak up water like a sponge.

Indicators of development
Statistics/figures used to determine how developed a country is. Classified as social, economic or composite indicators.

Karst scenery
Areas of mainly carboniferous limestone affected by chemical weathering by acidic rainfall.

Latitude
Lines running parallel to the Equator.

Levees
Naturally formed embankments found along the sides of rivers in their middle and lower courses.

Longshore drift
The movement of sediment along the coastline, caused by wave transportation.

Malaria
A disease spread by mosquitoes.

Meanders
Naturally occurring bends in rivers created by turbulent corkscrewing river action.

Migration
The movement of people either within a country or between countries, either voluntary or forced.

Misfit stream
A stream or river in a glaciated valley that is too small to have eroded the valley in which it flows.

Natural disaster
A major adverse event resulting from natural processes of the Earth; examples include floods, volcanic eruptions, earthquakes, tsunamis.

Névé
Old, compacted snow.

Glossary

Pedestrianisation
A street where cars are not permitted to go.

Physical quality of life index
An attempt to measure the quality of life or well-being of a country. The value is the average of three statistics: basic literacy rate, infant mortality, and life expectancy at age one, all equally weighted on a 0 to 100 scale.

Plucking
A process of glacial eroision by which ice freezes onto weathered rock and as it moves, pulls pieces of rock with it.

Pyramidal peak
A triangular shaped mouintain formed by three or more corries erode back to back.

Ranching
A type of farming that uses large areas of land to raise cattle, sheep and other livestock. The land is predominantly used for grazing and large numbers of livestock will be involved.

Regeneration
Improving a run-down urban area.

Relief
The shape of the Earth's surface.

Ribbon lake
A long narrow lake found on the floor of a glaciated valley.

Rural-urban fringe
A zone of transition between the built-up area and the countryside where there is often competition for land use.

Rural-urban migration
Movement of people from the countryside to cities.

Scarp
A steep slope often with a cliff that runs across the landscape and relates to erosion and exposure of underlying rock layers or to processes like faulting that involves uplift of the land.

Self-help schemes
Where groups of people, especially in developing countries, are encouraged to build or improve their own homes using materials provided by the local authority.

Solar radiation
Heat from the sun.

Sustainability
Land use systems that can be continued into the future without short, mid-term, or permanent damage to the environment and its ecosystems.

Tourism
A global industry that involves short medium and long-stay journeys to areas beyond home.

Transect
A series of zones that transition from sparse rural farmhouses to the dense urban core.

Transition zone
An area of change within a city where the boundaries between zones is blurred (e.g. the move from the C.B.D. to the inner city).

Transportation
The movement of materials by agents of erosion such as running water, glaciers, the wind and the sea.

Truncated spur
A former interlocking spur in a pre-glacial V-shaped valley that had its end removed by a glacier.

U-shaped valley
A wide valley with steep sides and a flat floor formed by glacial erosion.

V-shaped valley
A narrow, steep-sided valley formed by the rapid vertical erosion of a river.

National 5
GEOGRAPHY

Practice Papers

Fiona Williamson

Introduction

Layout of the practice papers

These practice test papers mirror the actual SQA exam as much as possible. The layout, paper colour and question level are similar to the actual exam that you will sit, so that you are familiar with what the exam paper will look like.

The answer section is provided online at www.collins.co.uk/pages/Scottish-curriculum-free-resources. The answers include practical tips on how to tackle certain types of questions, details of how marks are awarded and advice on what the examiners will be looking for. You will also find that we have included some additional answers to help you with your revision, e.g. the formation of stalactites and stalagmites is given alongside the answer for limestone pavement in question 2b in Exam B.

How To Use This Book

The practice papers can be used in two main ways:

1. You can complete an entire practice paper as preparation for the final exam. If you would like to use the book in this way, you can complete the practice paper under exam style conditions by setting yourself a time for each paper and answering it as well as possible without using any references or notes.

2. Alternatively, you can answer the practice paper questions as a revision exercise, using your notes to produce a model answer. Your teacher may mark these for you.

The Exam

Listed below is an outline of the topics assessed in the National 5 Geography Exam. You can obtain a detailed course outline from the SQA website at sqa.org.uk or ask your teacher for a copy.

The exam is divided into three sections:

1. Physical Environments

*   Weather – the effect of latitude, relief, aspect and distance from sea on local weather conditions; the characteristics of the five main air masses affecting the UK; the characteristics of weather associated with depressions and anticyclones

- Landscapes formation — **either** Glaciated Uplands and Coastal Landscapes **or** Upland Limestone and Rivers and Valleys

- Land uses including farming, forestry, industry, recreation and tourism, water storage and supply and renewable energy

- Land use conflicts and solutions

2. Human Environments

In relation to developing and developed countries

- Indicators of development

- Population distribution

- Birth rates and death rates

In relation to urban areas

- Land use zones in cities in the developed world

- Recent developments in the CBD, inner city, rural/urban fringe in developed world cities

- Issues/Solutions in developing world cities

In relation to rural areas

- Changes/modern developments to farming in developed countries

- Changes/modern developments to farming in developing countries

3. Global Issues

Candidates should study at least **two** global issues from the following:

Climate change

- Features of climate change

- Physical and human causes of climate change

- Local and global effects

- Managing and reducing its impact/effects

Natural regions

- Tundra and equatorial climates and their ecosystems

- Uses and misuses of these areas

- Effects of degradation on people and the environment

- Managing and reducing its impact/effects

Environmental hazards

- The main features of earthquakes, volcanoes and tropical storms

- Causes of each hazard

- Impact on the landscape and population of each hazard

- Management – methods of prediction, planning and strategies adopted in response to environmental hazards

Trade and globalisation

- Patterns of world trade

- Cause of inequalities in trade

- Impact of world trade patterns on people and the environment

- Methods to reduce inequalities — trade alliances, fair trade, sustainable practices

Tourism

- Mass tourism and eco-tourism

- Causes of/reasons for mass tourism and eco-tourism

- Effect of mass tourism and eco-tourism on people and the environment

- Methods adopted to manage tourism

Health

- Distribution of selected world diseases

- Causes, effects and methods adopted to manage:
  - AIDS in developed and developing countries
  - one disease prevalent in a developed country (choose from: heart disease, cancer, asthma)
  - one disease prevalent in a developing country (choose from: malaria, cholera, kwashiorkor, pneumonia)

Skills

Skills will be assessed across the three areas. Ordnance Survey (OS) mapping skills will be assessed in the Physical Environments and Human Environments units. Handling, extracting and interpreting information from graphs, tables and statistics will be assessed in all three units.

General Information about the exam

You have 2 hours 20 minutes to complete the paper.

There are three sections worth 80 marks in total.

Section 1 — Physical Environments — 30 marks

In this section you have a choice!

Attempt **EITHER** question 1 **OR** question 2, **AND** all others.

You will have studied two landscape types, **either** Glaciated Uplands and Coastal Landscapes **or** Upland Limestone and Rivers and Valleys. Make sure you choose the correct topic you have studied. Do not attempt to answer both question 1 and 2. **If you answer question 1 do not answer question 2.**

Section 2 — Human Environments — 30 marks

Attempt all the questions in this section.

Section 3 — Global Issues — 20 marks

In this section you have a choice!

There are six questions. You should attempt any **TWO** of the following; **do not answer all the questions.**

Climate Change

Natural Regions

Environmental Hazards

Trade and Globalisation

Tourism

Health

Make sure you choose the topics you have studied!

The Exam Paper

Always check you have the correct paper before you begin the exam.

Complete your exam in either blue or black ink. Always make sure you have a spare pen in case one runs out. A pencil is useful for diagrams.

Answering the Questions

You can answer the questions in whatever order you feel comfortable with but remember to clearly identify the question number you are attempting.

Carefully read each question so you answer what the question is asking. Put as much detail into your answer as possible – detail will gain you marks.

In questions that ask you to explain the formation of a feature, e.g. a corrie, it is possible to gain full marks by producing a series of diagrams. To do this you should add labels with explanations for your diagrams.

Do not use bullet points or lists, you should answer in sentences. The examiner is looking for you to show your knowledge of the course topics and a simple list is too basic and will not gain you full marks for the question.

Always attempt a question. Even if you are not sure of the answer you may pick up a few marks for your attempt.

There will be Ordnance Survey map questions. These questions could come up in the Physical section, the Human section, or both.

There will probably be more than one Ordnance Survey map. Maps will have item letters, e.g. Item A, Item B etc. The question will clearly state which map to use by indicating the item letter. There may be a different map used in the Physical section from the Human section – make sure you select the correct map. In Ordnance Survey questions always give evidence from the map, e.g. if you are asked for evidence to show the CBD of an area don't just use your knowledge to state the features of a CBD – you need to give actual examples from the Ordnance Survey map. Also, at National 5 you will be expected to use six figure grid references, although in some cases four figures will be sufficient.

There are six types of question used in these practice papers.

1. Describe	This is basically a list of facts. No reasons are needed in your answer.
	Example: **Describe** the location of an out of town shopping centre.
	Answer: It is found on the edge of town. There is plenty of housing close by. There are large areas of flat land with roads nearby.
2. Explain and **3.** Give reasons	This means you have to give reasons to support your answer using words like 'because', 'which means that', 'since' and 'as'.
	Below is the same question but answered as an explanation.
	Example: **Explain** the location of an out of town shopping centre. *Answer*: It is found on the outskirts of a town **because** there is plenty of room to expand. There is housing close by so that **means that** there are customers to buy the goods from the shopping centre. The housing **means that** there will be a labour force available for the shopping centre. The areas of flat land **means** it is easier to build on and there is plenty of land for car parking **as** most customers will arrive by car. There are roads nearby **which means that** goods can be transported in and out. **Since** more people will arrive by car the roads nearby make the area more accessible.

4. Match	In this type of question you will be asked to use your Ordnance Survey skills to match grid references to specific items.
	Example: **Match** the following grid references to the correct feature
	Grid References 367198, 673056, 234875
	Choose from: forestry, quarry, loch, A124
5. Give map evidence	This means you have to use map evidence in your answer. You need to pick out specific features from the map.
	Example: **Give map evidence** to show that the CBD of Dunfermline is found in grid square 378695
	Answer: There are many churches. There are museums. There is a town hall. There is tourist information. All the transport routes meet in this square.
6. Give advantages and disadvantages	In this type of question you need to evaluate the good and bad points of a particular situation.
	Example: **Give the advantages and disadvantages** of deforestation.
	Answer: Advantages: The trees can be exported to make money for the country. The extraction of timber provides jobs for the people. The money for the trees can be used to improve the lives of the local people.
	Disadvantages: The trees are the habitat of many species of birds and animals which could become extinct. The native tribes are forced from their homes and their traditional way of life threatened. Soil erosion occurs as there are no tree roots to bind the soil together.

Timing is important. The Physical and the Human sections are worth 30 marks each so you should aim to complete each section in about 50 minutes. The Global Issues section is worth 20 marks so you should aim to complete this in about 20 minutes. This should give you time to read the questions, make your selection in the choice section and leave you enough time to look over your paper at the end. Do not leave the exam before the end. It should take the whole time allocated to complete your paper. If you have finished well before the end of the exam then you need to go back and put more detail into your answers or attempt unanswered questions.

Your Course award

The National 5 Course has two parts – a question paper (exam) and an assignment.

The question paper will be worth 80 marks and the assignment will have 20 marks. Your mark for the exam and the assignment will be combined to give you a mark out of 100. The question paper is therefore worth 80% of the overall marks for the Course assessment and the assignment is worth 20%.

Course assessment will provide the basis for grading attainment in the Course award.

The Course assessment is graded A–D. The grade is determined on the basis of the total mark for all Course assessments together.

Good luck!

Practice paper A

Practice paper A

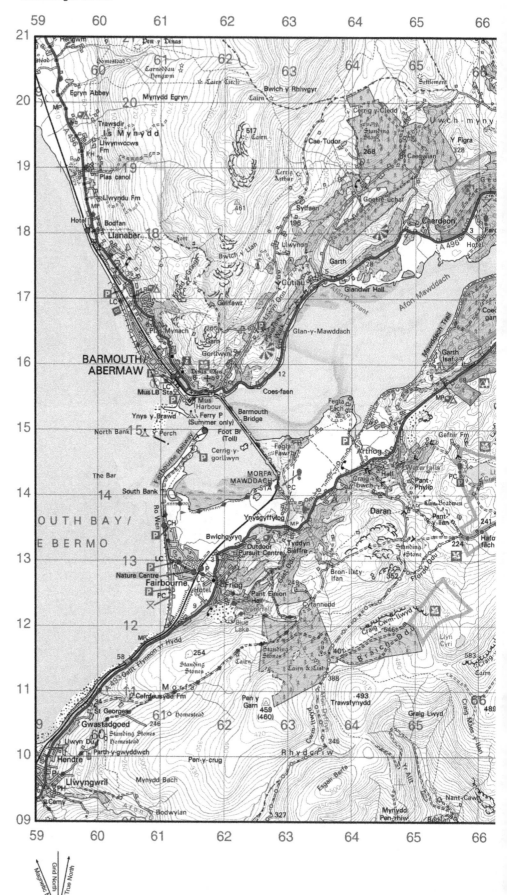

Grid North
True North
Magnetic North

Diagrammatic only

Ordnance Survey map of Barmouth

Four colours should appear above; if not then please return to the invigilator.

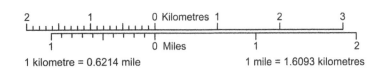

1 kilometre = 0.6214 mile   1 mile = 1.6093 kilometres

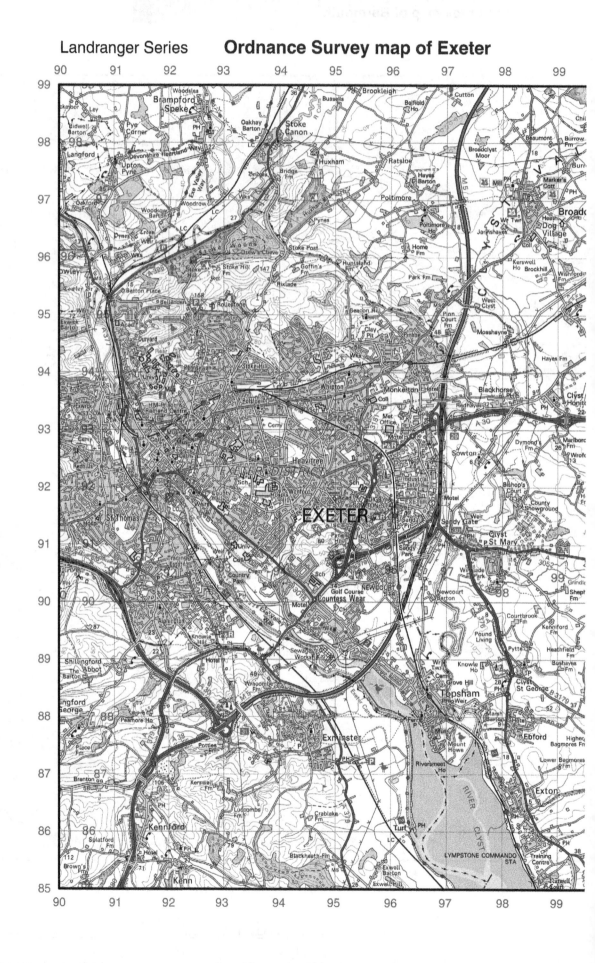

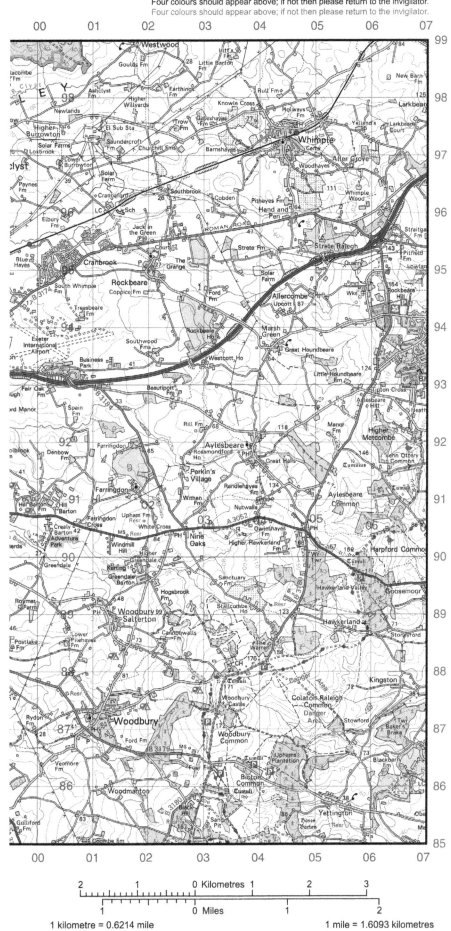

1 kilometre = 0.6214 mile   1 mile = 1.6093 kilometres

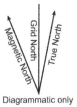

Diagrammatic only

EXAM A

Total marks — 80

Duration — 2 hours 20 minutes

SECTION 1 — PHYSICAL ENVIRONMENTS — 30 marks

In this section you must answer: **EITHER** Question 1 **OR** Question 2, **AND** Questions 3, 4 and 5.

SECTION 2 — HUMAN ENVIRONMENTS — 30 marks

In this section you must answer all questions.

SECTION 3 — GLOBAL ISSUES — 20 marks

Attempt any **TWO** questions

Climate Change

Natural Regions

Environmental Hazards

Trade and Globalisation

Tourism

Health

SECTION 1 — PHYSICAL ENVIRONMENTS

In this section you must answer **EITHER** Question 1 **OR** Question 2, **AND** Questions 3, 4 and 5

Question 1: Glaciated Uplands

Study the Ordnance Survey map of the Barmouth area.

a) Using grid references, **describe** the evidence shown on the map which suggests that this is an area of upland glaciation.   **4**

Diagram Q1: Formation of a Corrie

Before   After

b) Look at Diagram Q1.

   Explain the processes involved in the formation of a corrie.

   You may use diagram(s) in your answer.   **4**

NOW ANSWER QUESTIONS 3, 4 AND 5

MARKS

**DO NOT ANSWER THIS QUESTION IF YOU HAVE ALREADY
ANSWERED QUESTION 1**

Question 2: Rivers and Valleys

a) Describe the physical features of the Afon Mawddach and its valley between 729209 and
    695185.   **4**

Diagram Q2: Formation of a Waterfall

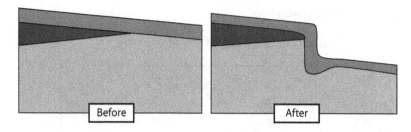

b) **Explain** the processes involved in the formation of a waterfall.

    You may wish to use a diagram(s) in your answer.   **4**

NOW ANSWER QUESTIONS 3, 4 AND 5

NOW ANSWER QUESTIONS 3, 4 and 5

Question 3

Diagram Q3: The Mawddach Way, Barmouth

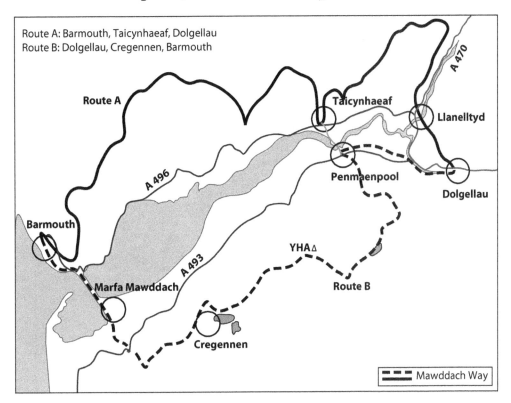

Study Diagram Q3 and OS map Item A of the Barmouth area.

A group of walkers have the choice of walking Route A or Route B of the Mawddach Way.

For either Route A or Route B, **describe the advantages and disadvantages** of the route. You must use map evidence in your answer.

6

MARKS

Question 4

Diagram Q4: Weather Chart UK, August

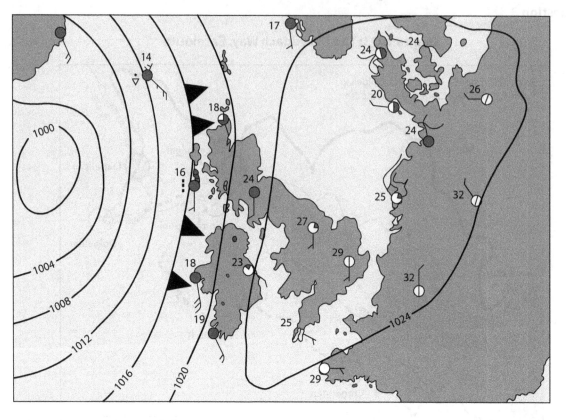

Study Diagram Q4.

a) **Describe** the differences in the weather conditions between the South of England and the North of Scotland. **4**

b) **Explain** the differences in weather conditions between the south of England and the north of Scotland. **6**

Question 5

Diagram Q5: Landscape Types and Land Uses

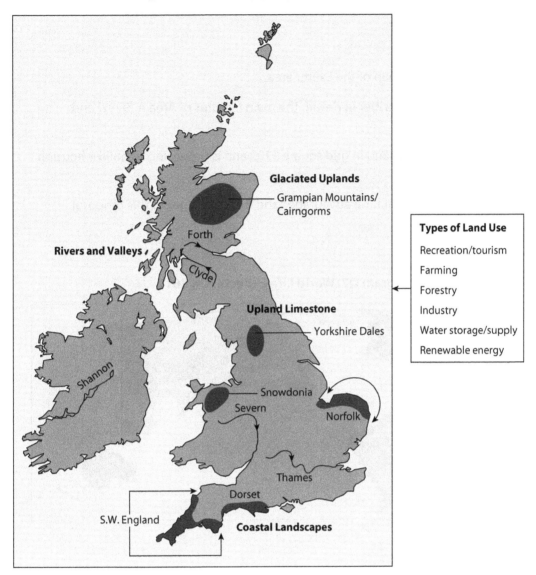

Look at Diagram Q5.

Choose **one** of the landscape types from Diagram Q5 which you have studied.

Explain the conflicts which can arise between any two land uses found in your chosen landscape. **6**

SECTION 2 — HUMAN ENVIRONMENTS

In this section you must answer Questions 6, 7, 8 and 9

Question 6

Study the Ordnance Survey map of the Exeter area.

a) Using map evidence, **describe, in detail,** the main features of Area A (9292) and
    Area B (9494).   **5**

b) A builder has bought the land in grid square 9792 and is proposing to build a housing
    estate in this square.

    Using map evidence explain the advantages and disadvantages of this proposal.   **5**

Question 7

Diagram Q7: World Life Expectancy, 2012

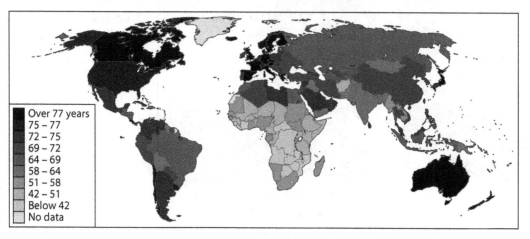

Over 77 years
75 – 77
72 – 75
69 – 72
64 – 69
58 – 64
51 – 58
42 – 51
Below 42
No data

Look at Diagram Q7.

Explain the reasons for the differences in life expectancy between developed and developing
countries.   **6**

Question 8

Diagram Q8: Population Pyramids for Scotland

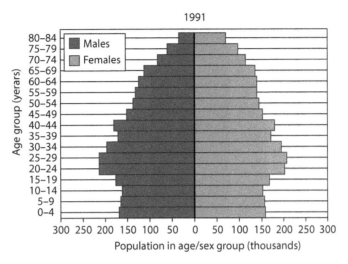

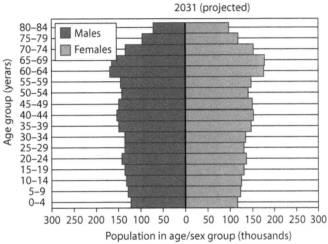

Study Diagram Q8.

Describe, in detail, the differences in Scotland's population structure between 1991 and 2031 (projected).   **4**

Question 9

For a named city in the developing world, **describe, in detail,** measures taken to improve conditions in shanty towns.   **4**

Question 10

Diagram Q10: Developments on the Rural/Urban Fringe in the UK

Out of town shopping centres

Housing

Airport expansion

Retail parks

Hotels and conference centres

Look at Diagram Q10.

Explain the attractions of the rural/urban fringe in the UK for developments like those shown in the diagram.

6

SECTION 3 — GLOBAL ISSUES

Answer any TWO questions

Question 11 — Climate Change

Question 12 — Natural Regions

Question 13 — Environmental Hazards

Question 14 — Trade and Globalisation

Question 15 — Tourism

Question 16 — Health

Question 11: Climate Change

Diagram Q11A: Changes in Global Mean Temperatures, 1880–2010

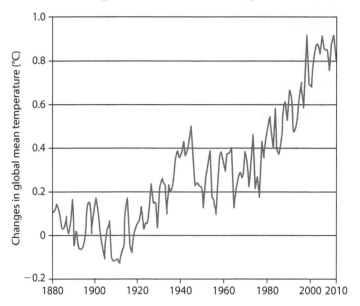

a) Study Diagram Q11A.

   Describe, in detail, the changes in global mean temperatures between 1880 and 2010.   **4**

Diagram Q11B: Effects of Global Warming

b) Look at Diagram Q11B.

   Explain the possible effects of global warming on the environment.   **6**

   You should refer to examples you have studied in your answer.

Question 12: Natural Regions

Diagram Q12: Climate Graphs

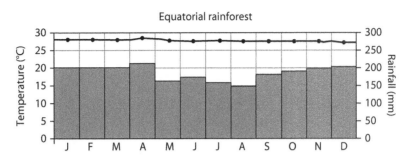

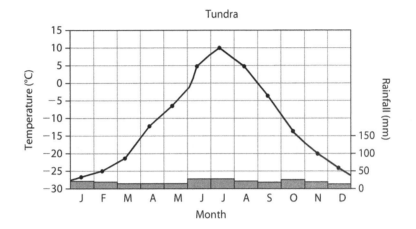

a) Study Diagram Q12.

Describe, in detail, the differences between the climates of the rainforest and tundra shown in the diagram above.

4

b) **Explain, in detail,** strategies used to reduce the impact of deforestation on people and the environment.

6

Question 13: Environmental Hazards

Diagram Q13: Distribution of Tropical Storms

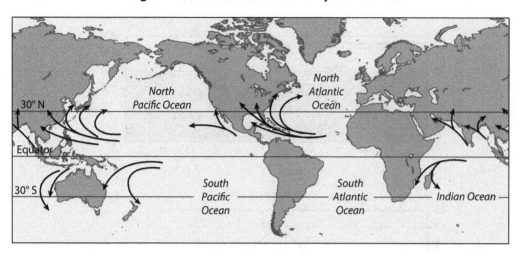

a) Study Diagram Q13.

   Describe, in detail, the distribution of tropical storms.   **4**

b) For a named tropical storm you have studied, **explain** the impact of the storm on
   the landscape and the population.   **6**

Question 14: Trade and Globalisation

Diagram Q14A: Growth in Volume of World Trade and GDP, 2005–2013

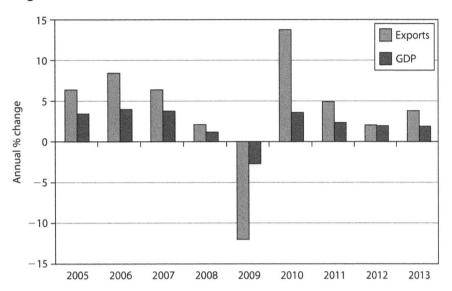

a) Study Diagram Q14A.

Describe, in detail, the changes in the growth of world trade and GDP between 2005 and 2013.

4

Diagram Q14B: Share of World Trade

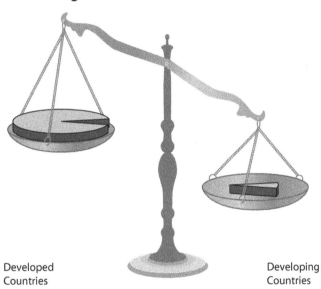

Developed Countries

Developing Countries

b) Look at diagram Q14B.

Explain the reasons for the inequalities in world trade.

6

Question 15: Tourism

Diagram Q15A: Some Facts on Tourism in California

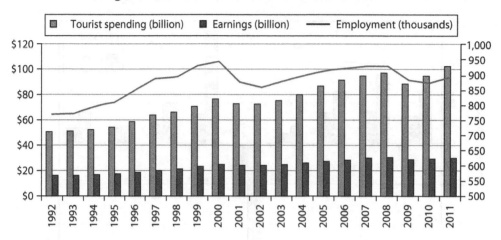

a) Study Diagram Q15A.

Describe, in detail, the changes in Californian tourism between 1992 and 2011.   **4**

Diagram Q15B: Newspaper Quote

> "The United Nations World Travel Organization estimates that in 2012 eco-tourism captured 11% of the international market"

b) Look at the statement above.

Explain the main features of eco-tourism.   **6**

Question 16: Health

Diagram Q16A: Distribution of Child Cases of Pneumonia

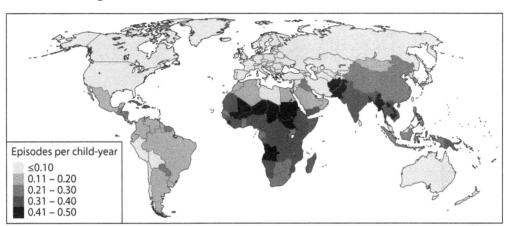

Episodes per child-year
≤0.10
0.11 – 0.20
0.21 – 0.30
0.31 – 0.40
0.41 – 0.50

a) Study Diagram Q16A.

Describe, in detail, the distribution of child cases of pneumonia. **4**

Diagram Q16B: AIDS Facts

More than 36 million people live with AIDS/HIV

In 2012, 1.9 million people died from AIDS

Every day nearly 7,000 people contract HIV

b) Look at Diagram Q16B.

Explain the causes of AIDS and some strategies used to control/manage the disease. **6**

[END OF QUESTION PAPER]

Practice paper B

Practice paper B

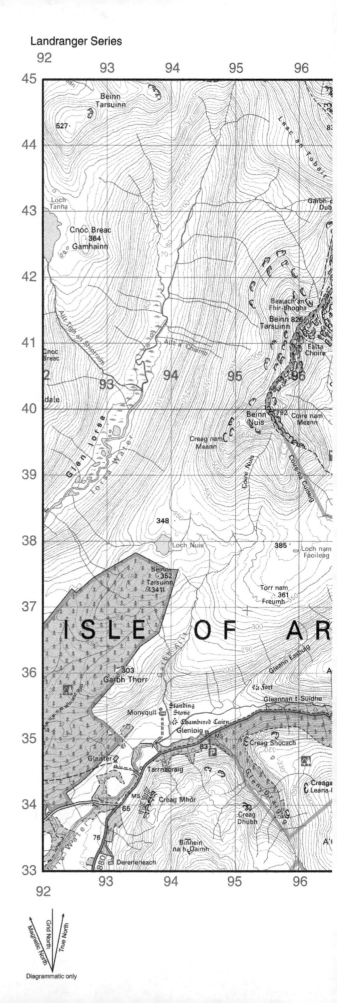

Ordnance Survey map of Arran

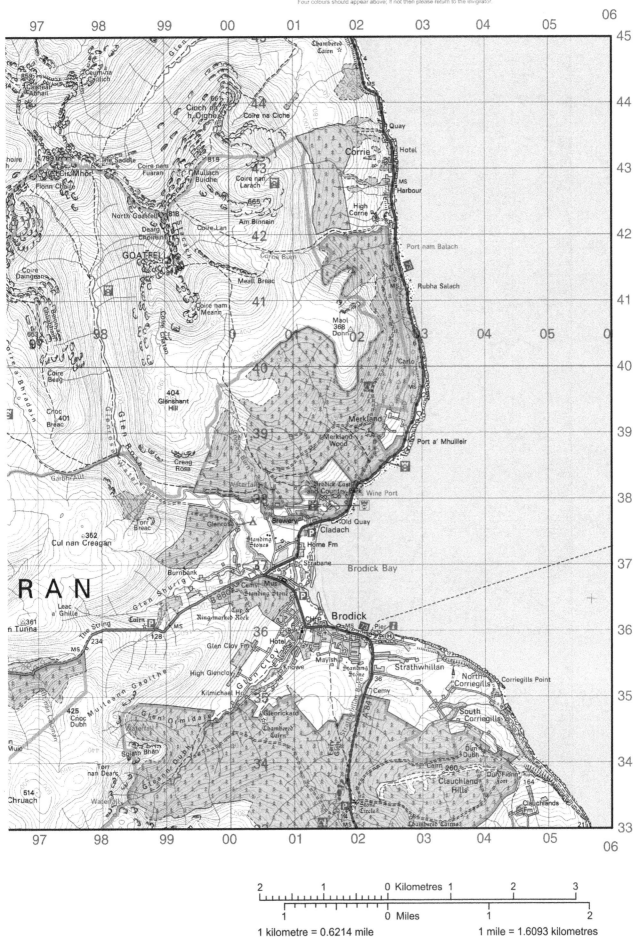

1 kilometre = 0.6214 mile   1 mile = 1.6093 kilometres

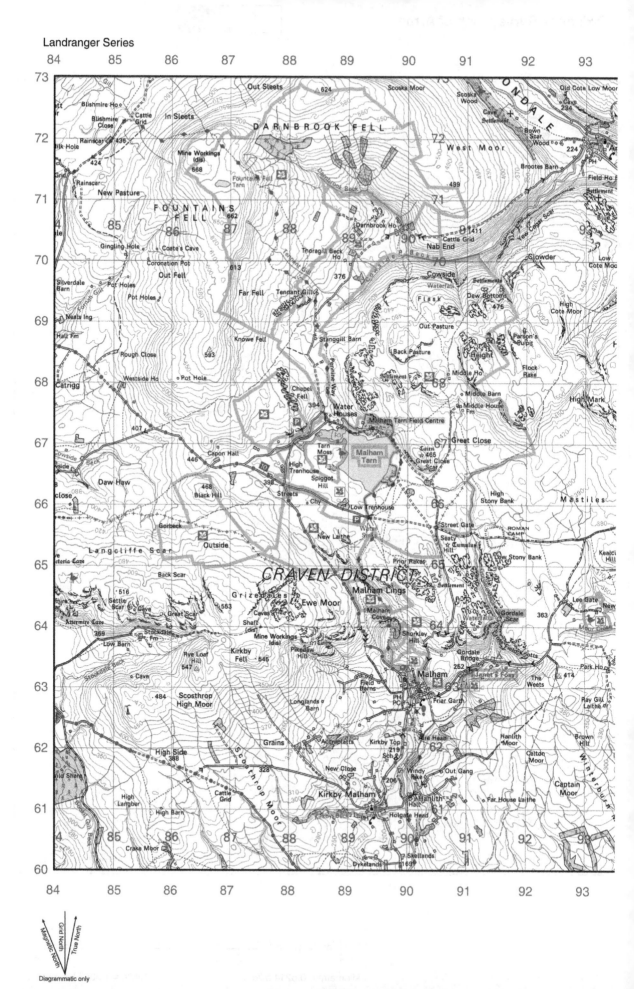

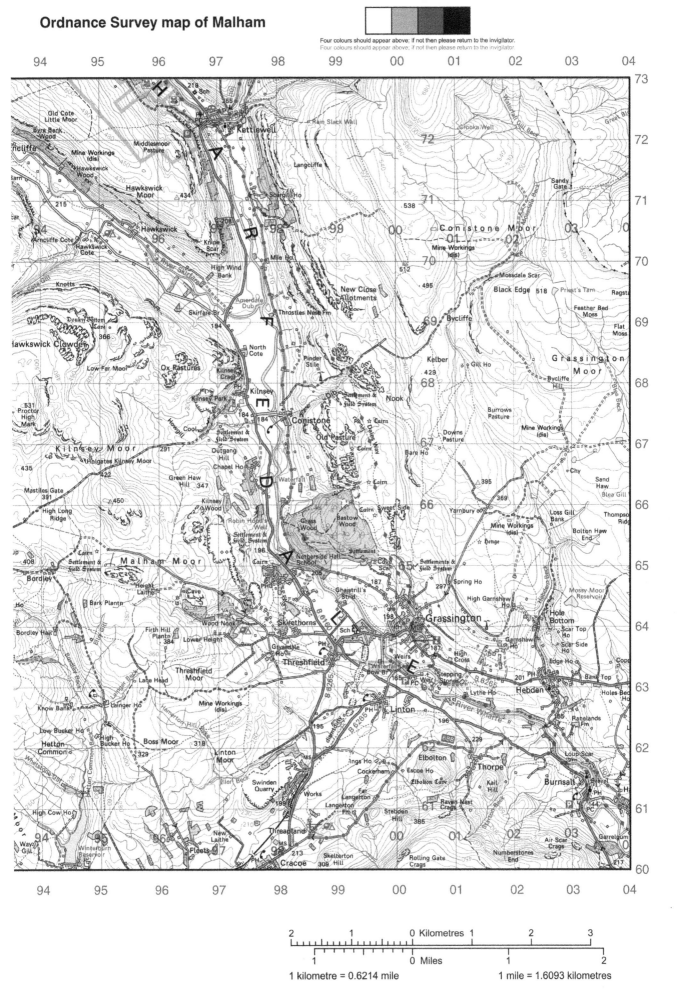

Ordnance Survey map of Malham

Four colours should appear above; if not then please return to the invigilator.
Four colours should appear above; if not then please return to the invigilator.

2   1   0 Kilometres 1   2   3

1   0 Miles   1   2

1 kilometre = 0.6214 mile   1 mile = 1.6093 kilometres

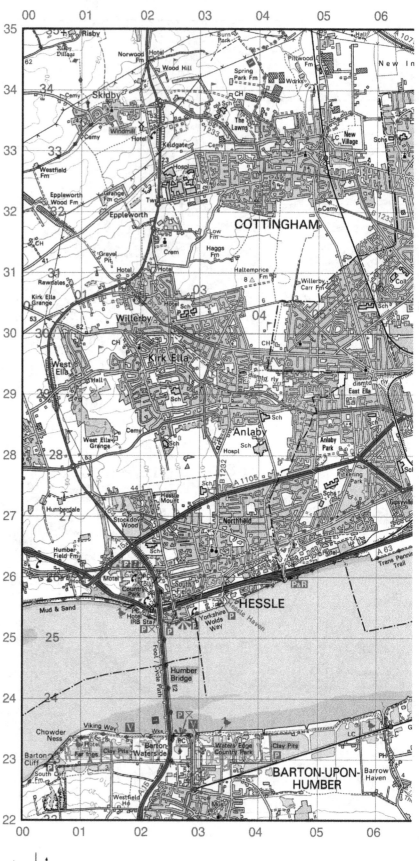

Grid North
True North
Magnetic North

Diagrammatic only

Ordnance Survey map of Hull

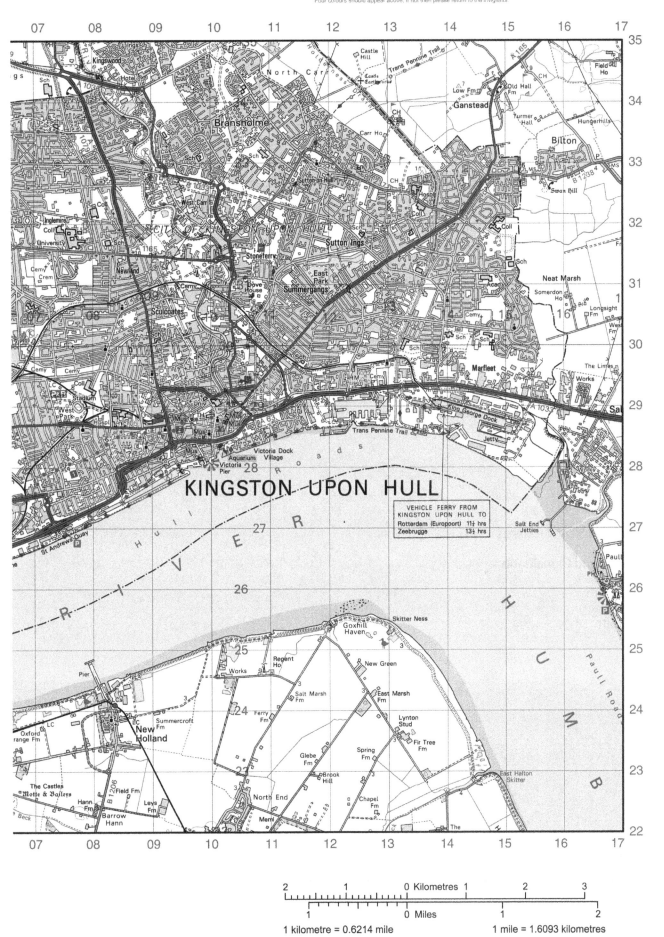

Four colours should appear above; if not then please return to the invigilator.
Four colours should appear above; if not then please return to the invigilator.

VEHICLE FERRY FROM
KINGSTON UPON HULL TO
Rotterdam (Europoort) 11¼ hrs
Zeebrugge 13½ hrs

KINGSTON UPON HULL

1 kilometre = 0.6214 mile   1 mile = 1.6093 kilometres

EXAM B

Total marks — 80

Duration — 2 hours 20 minutes

SECTION 1 — PHYSICAL ENVIRONMENTS — 30 marks

In this section you must answer: **EITHER** Question 1 **OR** Question 2, **AND** Questions 3, 4, 5, 6 and 7.

SECTION 2 — HUMAN ENVIRONMENTS — 30 marks

In this section you must answer all questions.

SECTION 3 — GLOBAL ISSUES — 20 marks

Attempt any **TWO** questions

Climate Change

Natural Regions

Environmental Hazards

Trade and Globalisation

Tourism

Health

SECTION 1 — PHYSICAL ENVIRONMENTS

In this section you must answer **EITHER** Question 1 **OR** Question 2, **AND** Questions 3, 4, 5, 6 and 7.

Question 1: Glaciated Uplands

Study the Ordnance Survey map of the Arran area.

a) Match the glaciated uplands features shown below with the correct grid reference.

pyramidal peak, arête, corrie

992416, 998438, 995409, 981401   **3**

b) **Explain** the formation of an arête.

You may use a diagram(s) in your answer.   **4**

NOW ANSWER QUESTIONS 3, 4, 5, 6 and 7

DO NOT ANSWER THIS QUESTION IF YOU HAVE ALREADY ANSWERED QUESTION 1

Question 2: Upland Limestone

Study the Ordnance Survey map of the Malham area.

a) Use the information in the Ordnance Survey map to match the features of upland limestone below with the correct grid reference.

pothole, limestone pavement, disappearing, stream

902647, 861681, 870660, 894657   3

b) **Explain** the formation of a limestone pavement.

You may use a diagram(s) in your answer.   4

NOW ANSWER QUESTIONS 3, 4, 5, 6 and 7

NOW ANSWER QUESTIONS 3, 4, 5, 6 and 7

Question 3

Study the Ordnance Survey map of the Arran area

Explain ways in which the physical landscape has affected land use on the map extract.

5

Question 4

Diagram Q4: Synoptic Chart, 5 December 2013

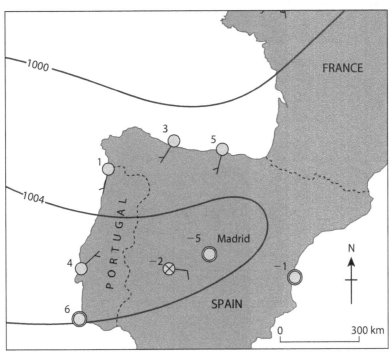

Study Diagram Q4.

Explain the weather conditions being experienced in Spain.

4

Question 5

Explain the advantages and disadvantages that a Tropical Continental air mass brings to the people of the UK in summer.

4

Question 6

Diagram Q6A: Landscape Types in the UK

| Glaciated Uplands | Upland Limestone | Coastal Landscapes | Rivers and Valleys |

Diagram Q6B: Land Uses in the UK

Forestry

Water storage and supply

Industry   Farming

Recreation and tourism

Renewable energy

Look at Diagrams Q6A and Q6B

Choose **one** landscape type you have studied from Diagram Q6A.

Select **two** land uses from Diagram Q6B.

Explain why your selected land uses are suitable for your chosen landscape type.   **5**

Question 7

"A land use conflict occurs when there are conflicting views about how the land should be used".

Referring to a named area you have studied, **explain** how land use conflicts can be managed.   **5**

SECTION 2 — HUMAN ENVIRONMENTS

In this section you must answer Questions 8, 9, 10 and 11

Question 8

Study the Ordnance Survey map of the Kingston upon Hull area.

a) **Use map evidence** to show that part of Kingston upon Hull's CBD is found in grid square 0928.   **3**

b) Using map evidence **describe, in detail**, the differences in the urban environments between grid square 0727 and grid square 0228.   **5**

c) Mr and Mrs Bennett and their teenage son are moving to Kingston Upon Hull. They have decided to buy a house in either grid square 0929 or grid square 0222. Which area should they choose? **Use map evidence** to support your choice.   **5**

Question 9

Diagram Q9: Development Indicators

Social Indicators	Economic Indicators
Life expectancy at birth	Gross domestic product (GDP)
Population per doctor	Energy used per person
Percentage adult literacy	Percentage population employed in agriculture

Study Diagram Q9.

Choose one social **and** one economic indicator.

Explain how your chosen indicators can show the level of development of a country.   **4**

Question 10

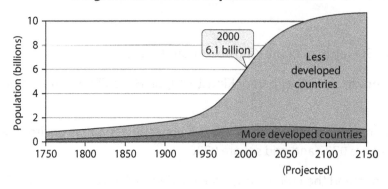

Diagram Q10: World Population Growth

a) Look at Diagram Q10.

Explain why the population of developing countries is growing far more rapidly than that of developed countries. **5**

b) **Describe** ways in which developing countries can reduce population growth. **3**

Question 11

Explain how recent developments in agriculture affect the people and environment of developed counties like the UK. **5**

SECTION 3 — GLOBAL ISSUES

Answer any TWO questions

Question 12 — Climate Change

Question 13 — Natural Regions

Question 14 — Environmental Hazards

Question 15 — Trade and Globalisation

Question 16 — Tourism

Question 17 — Health

Question 12: Climate Change

Diagram Q12A: Global Temperatures and Carbon Dioxide Emissions

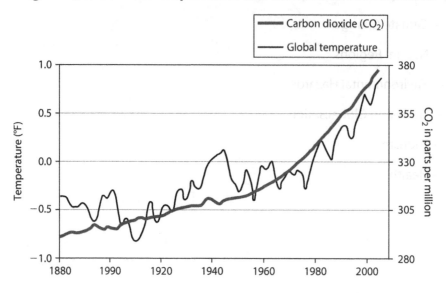

a) Study Diagram Q12A.

   Describe, **in detail**, the changes in global temperatures and carbon dioxide emissions.　　**4**

Diagram Q12B: Some Solutions to Global Warming

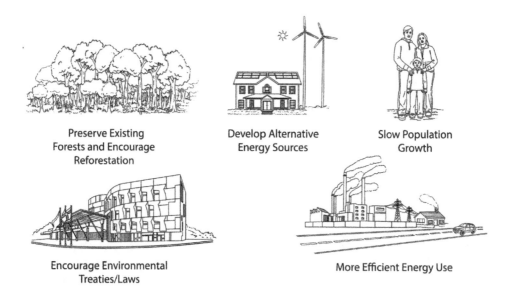

Preserve Existing Forests and Encourage Reforestation

Develop Alternative Energy Sources

Slow Population Growth

Encourage Environmental Treaties/Laws

More Efficient Energy Use

b) Look at Diagram Q12B.

   Choose **two** solutions from the diagram.

   Explain how each chosen solution could reduce global warming.　　**6**

Question 13: Natural Regions

Diagram Q13: Deforestation in Brazil, Bolivia and Peru

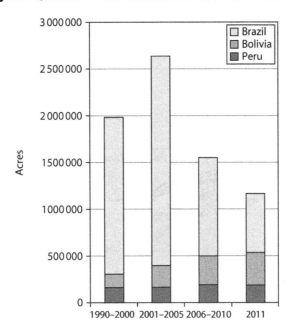

a) Study Diagram Q13.

Describe, in detail, the changes in deforestation in Brazil, Bolivia and Peru. **4**

b) **Explain** the benefits and problems of developing **either** the rainforest **or** the tundra. **6**

Question 14: Environmental Hazards

Diagram Q14A: Track of Selected July Tropical Storms

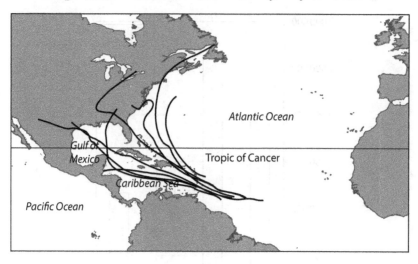

a) Study Diagram Q14A.

   Describe, in detail, the tracks of the tropical storms shown on Diagram Q14A.   4

Diagram Q14B: Areas hit by Hurricane Katrina

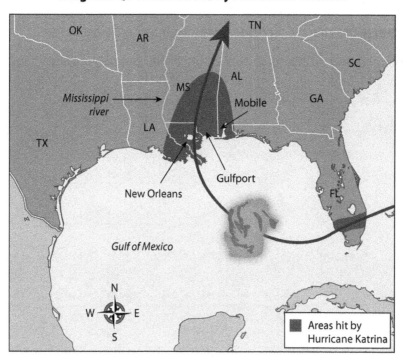

b) Look at Diagram Q14B.

   For a named hurricane, earthquake or volcanic eruption you have studied, **explain** the impact on the people and environment of the area.   6

Question 15: Trade and Globalisation

Diagram Q15A: UK's Export Trade with Selected Regions, 1999–2030 (projected)

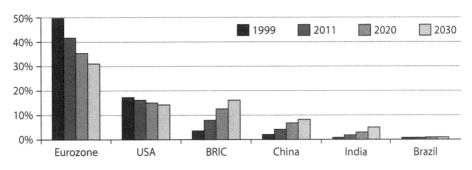

a) Study Diagram Q15A.

Describe, in detail, the projected changes in the UK's export trade between 1999 and 2030.　　　　　　　　　　　　　　　　　　　　　　　**4**

Diagram Q15B: Strategies to Protect Developing World Economies

b) Look at Diagram Q15B.

Choose either fair trade or cooperatives.

Explain how strategies like fair trade and cooperatives can protect the economy, people and environment of a developing world economy.　　　　　　　**6**

Question 16: Tourism

Diagram Q16: Indian Tourism 1995–2012

Year	Tourist arrivals to India (millions)	India's share of world tourism (%)	Earnings from tourism (US $ billion)
1995	2.12	0.39	2.58
2000	2.65	0.39	3.46
2002	2.38	0.34	2.92
2004	3.46	0.45	6.17
2006	4.45	0.53	8.63
2008	5.28	0.57	11.75
2010	5.58	1.12	14.19
2012	6.66	1.29	15.95

a) Study Diagram Q16.

Describe, in detail, the trends in Indian tourism between 1995 and 2012. **4**

b) For any named area you have studied, **explain ways** in which the effects of tourism can be controlled and managed. **6**

Question 17: Health

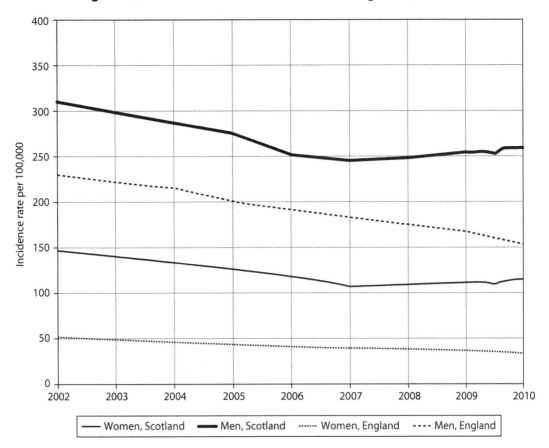

Diagram Q17: Heart Attacks in Scotland and England (2002–2010)

a) Study Diagram Q17.

   Describe, in detail, the differences in numbers of heart attacks between Scotland
   and England.

4

b) For **either** malaria, cholera, heart disease **or** cancer **explain** the main causes of the disease
   and describe some methods used to combat the disease.

6

[END OF QUESTION PAPER]

How your paper will be marked

In this section you have the opportunity to see how questions are marked by the marker.

One question has been taken from each of the practice papers in this book and two sample answers for each have been provided. Explanations of the marks achieved have been provided to help you identify the differences between a poor/good/excellent answer.

A marker needs to follow marking instructions and takes the following into account:

- Knowledge of the subject – has the candidate answered the question?

- If a question asks for explanation then the marker will allocate marks to statements that give reasons but not to simple descriptions (see types of question on page 7).

- If a question asks for two different points to be discussed then for the marker to allocate full marks both points must be mentioned. For example, if a question is marked out of six and asks for human and physical factors then both need to be mentioned for full marks. Marks are not necessarily allocated 3:3. For example, four marks could be allocated to one factor and two to the other, or five marks to one factor and one mark to the other. This allocation varies according to the marking instructions given to markers.

- If the question asks for differences then that needs to be made clear to the marker in the answer, e.g. "it is cloudy in Scotland **whereas** it's sunny in England".

- If map evidence is asked for then the marker is looking for reference to the specific map being used.

- In 'skills' questions, basic facts should be given based on the graph/diagram used. Quote dates, directions, figures or trends.

- Information lifted straight from a diagram will achieve very few marks, if any. The information needs to be processed/explained in some way.

Exam A, Question 16b (Global Issues)

For Malaria, Cholera, Pneumonia or Kwashiorkor, **explain** the consequences of the disease for the population in an affected area.

You should refer to examples you have studied in your answer. **6**

Answer 1

Malaria is common in poor countries like Nigeria. The disease is caused by a mosquito biting a human. This causes the person to become ill and unable to work (✓). If the person takes lots of time off then they may lose their job (✓). In many poor countries, farming is the main industry and problems will be caused if a person cannot go to work (✓). People are sometimes forced to move away from their village to escape breeding mosquitoes (✓). Many babies will be affected as there is a lack of mosquito nets and health care (✓). If the mosquito is not killed the disease will keep on spreading.

Comments/Marking

In general, one valid point will gain you a mark. If an answer is worth 6 marks you need to make six points.

This is a good answer but with room for improvement. On the whole it makes relevant points on the effects on the people. However, the first two sentences and the last sentence are not effects but simple descriptions of where the disease occurs and information about the mosquito. These are not relevant to this question so gain no marks. The next sentences are appropriate explanations so gain marks. The statement on people not being able to work on farms is a repeat (R) of the previous point so gains no marks. Remember, the question asks for explanation so your answer should show how the people are affected. This answer would get 4 out 6 marks.

Answer 2

Nigeria is a country I have studied where 97% of the population is at risk from malaria. A consequence for people living in areas affected is that sufferers can no longer work (✓), not only that but other people also spend time looking after them which further decreases the workforce (✓). If too much time is taken off work then they may lose their job (✓), reducing the family income making it difficult to feed their families (✓) and causing an increase in poverty in general (✓), as well as decreasing the economy of the area (✓). Malaria is a major public health problem in Nigeria where it accounts for more cases and deaths than any other country in the world (✓). If not seen by a doctor, a person infected with malaria has a risk of not surviving (✓). Medicines and treatment are expensive but not being able to work means that the person finds it difficult to pay for medicines and treatment (✓) thus increasing the burden on their families (✓). People are sometimes forced to leave their homes and fields to get away from malaria infected areas and this can result in hardship as they lose their homes and their livelihood (✓).

Comments/Marking

This is an excellent answer which correctly identifies the consequences of living with malaria. It begins well by referring to an appropriate named area, Nigeria, and includes facts on Nigeria in the answer. This shows your depth of knowledge of a particular area to the marker. The question asks for a case study you have studied so, by inserting appropriate facts, this answer can achieve full marks. This answer would get full marks, 6 out of 6. In fact, sufficient points were made to gain 11 marks!

Exam B, Question 3 (Physical Environments)

Diagram Q3: The Mawddach Way, Barmouth

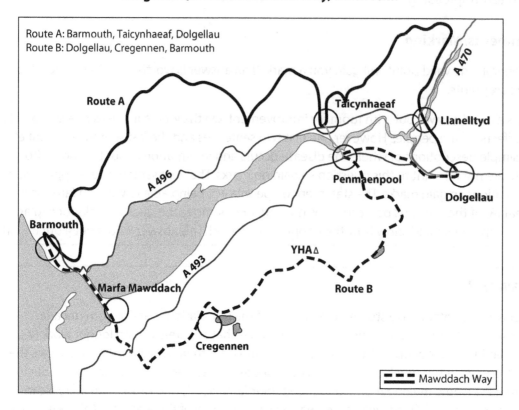

Route A: Barmouth, Taicynhaeaf, Dolgellau
Route B: Dolgellau, Cregennen, Barmouth

Study Diagram Q3 and OS map Item A of the Barmouth area.

A group of walkers have the choice of walking Route A or Route B of the Mawddach Way.

For either Route A or Route B, **describe the advantages and disadvantages** of the route. You must use map evidence in your answer.

6

Answer 1

I think they should choose Route A. It has plenty of parking at the start. It is close to the National Park with mountains and forests. There is a nature reserve at 687194 (✓) where a variety of wildlife can be seen (✓). There is a hotel where you could get a drink and food (✓). It is away from towns so will be a peaceful walk (✓) but there is a telephone box at 691194 where they could contact the police if there was an emergency (✓). The route could be difficult for children as it is quite high in places (✓) and it passes an old mine.

Comments/Marking

This is a good answer but takes a while to get to the point. The first three statements are descriptions. To gain marks, an advantage or disadvantage of the feature identified needs to be given, e.g. the forests can be used for picnicking or bird spotting. One mark is gained for giving an appropriate grid reference at 687194. Another correct grid reference is given for the telephone box but the mark has already been given for the nature reserve. The answer also concentrates more on the advantages of the walk, with only one mark being gained for the disadvantages. Passing an old mine needs to be expanded to include the point that it spoils the view. This answer would achieve full marks, 5 out of 5, four marks coming from the advantages and one mark from the disadvantages. If no disadvantages had been given this answer would achieve 4 marks.

Answer 2

My choice is Route B

The advantages of this route are that there is a variety of scenery along the walk (✓), from views of the coast, to mountains and rivers (✓). A variety of wildlife can be seen in these areas (✓), from sea birds on the coast to deer on the mountains (✓). There are places along the route to stop for something to eat such as the picnic site (✓) at 697153 (✓). There are no settlements in the area so there will be limited traffic, meaning the route will be quiet and peaceful (✓).

The disadvantages of the route are that it is quite high up so could be a difficult climb for small children (✓) and it could be dangerous going down as it is quite steep (✓) at 658145 (✓), At 240 metres it could be cold (✓) so walkers would need to carry warm clothing. There are very few places to stop along the route so getting something to eat and drink would be difficult (✓) and they would need to carry it with them making their packs heavy (✓).

Comment/Marking

This is a better answer. It is well organised. It separates the advantages from the disadvantages making it clearer for the marker. It uses map evidence to support each statement and gives an appropriate grid reference which gains one mark. As requested it gives a balance of advantages and disadvantages. Unlike the first answer it takes each statement made and backs it up with evidence from the map. There are seven marks available for the advantages and six marks for the disadvantages so easily gains full marks, 5 out of 5.

Exam C, Question 9 (Human Environments)

Diagram Q9: Development Indicators

Social Indicators	Economic Indicators
Life expectancy at birth	Gross domestic product (GDP)
Population per doctor	Energy used per person
Percentage adult literacy	Percentage population employed in agriculture

Study Diagram Q9.

Choose one social **and** one economic indicator.

Explain how your chosen indicators can show the level of development of a country.   4

Answer 1

I have chosen Population per doctor. This means the number of people in an area allocated to one doctor. In some areas of the world there are large numbers of people to one doctor while in others there are less people to one doctor. It is better to have a small number of people to one doctor.

My second indicator is Percentage population employed in agriculture. This is the number of people who work in farming. Farming is a primary industry and more people are employed in poor countries as they don't have many other jobs in industry (✓).

Comment/Answer

This is a poor answer. It does not answer the question. It is a description of the indicator not how that indicator can be used to tell about the development of a country. There is one mark at the end, although it's not too specific, as it implies there are few people employed in industry where more money is made so a country doesn't have money to develop. In this case the pupil made the correct choice of one economic and one social indicator. If the answer was related to only one type of indicator then only two marks would be available. This answer achieves only 1 out of 4.

Answer 2

The social indicator I have chosen is Population per doctor. This is the number of the people to one doctor. The lower the number of people per doctor the better, because countries with a low population per doctor are normally more developed which means they have a better health care system (✓) with more doctors trained meaning less patients per doctor (✓). This means that the country is also wealthy enough to be able to afford to invest money in a health care system (✓).

The economic indicator I have chosen is Percentage population employed in agriculture. A low percentage employed in agriculture usually shows a more developed country which means more people are employed in industry (✓). This means that people have a higher standard of living (✓) as these jobs pay better than jobs in farming (✓).

Comment/Marking

This is a very good answer. Two appropriate indicators have been discussed. The indicator has not just been described (which would in this case achieve no marks) but used to explain how it shows development. It makes three good points about each indicator achieving a possible six points for a four point answer. This answer achieves 4 out of 4.

Have a go yourself!

Marking is a good way to improve your exam technique and knowledge of the course. Mark your own answers to the practice papers following the hints and guidelines in this book. This will show you if your answers are too short, if they don't have enough detail in them, or don't answer the question. If you have a study partner you could mark each other's papers! You can then discuss/justify your marking. This will give you a better understanding of the questions and what is expected in an answer.

After you have completed the practice papers, have a go at marking the following answers (there are no marking schemes for these three questions). Mark the questions and then compare your marking with a study partner. Be prepared to explain why you did or did not give marks for a particular point!

Question 1

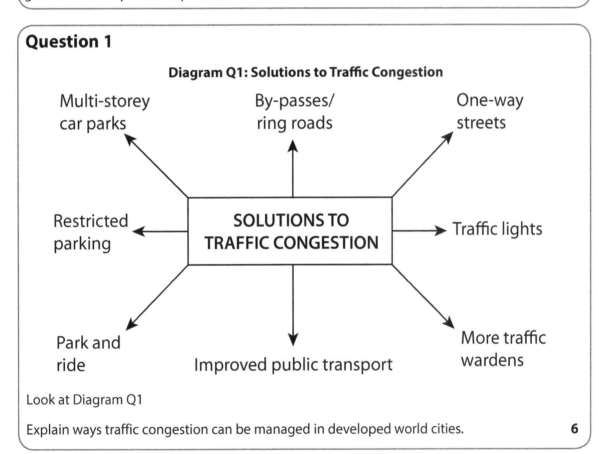

Diagram Q1: Solutions to Traffic Congestion

Look at Diagram Q1

Explain ways traffic congestion can be managed in developed world cities.   **6**

Answer

If you improve public transport more people will use buses, train, etc., and fewer people will use cars so the roads will become less congested. If fewer people use their cars it will also cause greenhouse gases to decrease. In addition, the transport company will make a profit – lots more people will use their services. However, buses and trains can be expensive so more people will use their cars rather than pay high prices. Also if the bus breaks down there will be more people let down and they may start to use their own cars again. Bypasses and ring roads allow traffic to go round the city instead of through which reduces the amount of cars in the city centre. Park and Ride schemes allow cars to be parked outside the city and a train can then be taken into the city. Trains carry more passengers thus reducing the cars entering the city.

Question 2

Diagram Q2: Birth rate for the UK and Chad

Country	Birth rate per 1000 of the population
United Kingdom	12
Chad	39

Look at Diagram Q2.

Give reasons to explain the differences in birth rates between a developed country such as the UK and a developing country such as Chad.   **4**

Answer

Reasons for the difference in birth rates between a developed country and a developing country are as follows. In developing countries, people depend on their children to help bring in money for their families and in olden days a lot of children used to die because they didn't have proper health care services. Families still think that their children will die so have a lot to make up for the ones they think they will lose. However, in a developed country, we don't depend on our children to help bring in money for the family and we know the chance of our children dying is less likely so our birth rate is lower. Developed countries that have more medical care have fewer children as there is more chance of them surviving.

Question 3

For a named natural disaster you have studied, explain methods use to reduce/manage the effects of a disaster.   **6**

Answer

In a hurricane, people are given warnings on the television days before it arrives to give them time to leave their homes and go to shelter somewhere safer. People can board up their houses and businesses to reduce damage. Some buildings are fitted with hurricane shutters to protect the property. Routes to take are signposted so that areas can be evacuated quickly. People get leaflets explaining what to do in a hurricane. People are warned of the danger of staying in their homes if they have been told to leave.